T0131043

# Neurological Effects of Repeated Exposure to Military Occupational Levels of Blast

## A Review of Scientific Literature

Molly M. Simmons, Charles C. Engel, Emily Hoch,
Patrick Orr, Brent Anderson, Gulrez Shah Azhar

Prepared for the United States Army

For more information on this publication, visit www.rand.org/t/RR2350

**Library of Congress Cataloging-in-Publication Data** is available for this publication.
ISBN: 978-1-9774-0292-9

Published by the RAND Corporation, Santa Monica, Calif.
© Copyright 2020 RAND Corporation
**RAND**® is a registered trademark.

*Cover photos (clockwise from top left): U.S. Army photo by Spec. Donald Williams;*
*U.S. Air Force photo by Staff Sgt. Evelyn Chavez; U.S. Marine Corps photo by Sgt. Mauricio Campino;*
*Department of Defense photo by Cpl. Tyler Main; U.S. Air Force photo; U.S. Army photo by Sgt. Michael Eaddy.*

**Support RAND**
Make a tax-deductible charitable contribution at
www.rand.org/giving/contribute

**www.rand.org**

# Preface

This report documents research and analysis conducted as part of a project entitled *Facilitating the Seventh Department of Defense State of the Science Meeting for Blast Injury Research*, sponsored by the U.S. Army Medical Research and Materiel Command. The purpose of the project was to facilitate the convening of the Seventh Department of Defense State-of-the-Science Meeting for Blast Injury.

This research was conducted within RAND Arroyo Center's Personnel, Training, and Health Program. RAND Arroyo Center, part of the RAND Corporation, is a federally funded research and development center (FFRDC) sponsored by the United States Army.

RAND operates under a "Federal-Wide Assurance" (FWA00003425) and complies with the *Code of Federal Regulations for the Protection of Human Subjects Under United States Law* (45 CFR 46), also known as "the Common Rule," as well as with the implementation guidance set forth in DoD Instruction 3216.02. As applicable, this compliance includes reviews and approvals by RAND's Institutional Review Board (the Human Subjects Protection Committee) and by the U.S. Army. The views of sources utilized in this study are solely their own and do not represent the official policy or position of DoD or the U.S. Government.

# Contents

# Figure and Tables

## Figure

## Tables

# Summary

This report details the state of the science regarding the relationship between occupational exposure to low-level blasts and nervous system problems in military service members. This literature review was completed as part of the 2017 State of the Science (SoS) Meeting sponsored by the United States Department of Defense (DoD) Blast Injury Research Program Coordinating Office. The goal of the SoS and its associated processes is to identify what is known and not known pertaining to key blast injury–related topics and emerging issues. The topic of the 7th SoS as well as this supporting literature review is "The Neurological Effects of Repeated Exposure to Military Occupational Levels of Blast: Implications for Health and Prevention."

Over the past decade, awareness of the central nervous system (CNS) effects of explosive blast exposure has increased. A key driver of that awareness has been the blast-related injuries suffered by service members during combat operations in Iraq and Afghanistan. As U.S. combat operations in these regions have largely come to a close, there has been growing concern over repetitive forms of blast exposure during military service that is, most often, unrelated to combat. Examples of these forms of exposure include exposure to weapon systems, such as "recoilless" rifles and shoulder-launched rocket launchers (e.g., the Carl Gustav antitank weapon), that can produce more than one blast exposure per round. The repeated exposure to these low-level blasts raises new questions of effects on CNS structure and function and on the health of those service members who have been exposed to repeated discharges. The literature review that follows has targeted the scientific literature pertaining to the effects of repeated military occupational blast exposures, specifically exploring what is known regarding the broad nervous system consequences of these exposures. For this review, *military occupational blast* (*MOB*) was defined as low-level repeated blast exposures that do not result in immediate *loss of consciousness* (*LoC*) and often may not even involve acute *alteration of consciousness* (*AoC*), and that have potential neurological health implications. Either LoC or AoC are prerequisites for mild traumatic brain injury (mTBI) and therefore, this report is focused mainly on sub-mTBI (i.e., subconcussive) blast exposure. The literature review addressed the following research questions:

1. What is known about the occurrence of repeated low-level occupational blast exposure incurred during military service?
2. What is the scientific evidence relating to the potential neurological health effects of repeated low-level occupational blast exposure?
3. What are promising strategies for preventing the potential neurological effects of repeated low-level military occupational blast exposure?
4. What are promising early detection indicators for the potential neurological consequences of repeated low-level military occupational blast exposure?

## Methods

We sought to review related studies to help us better understand short-, intermediate-, and long-term neurological outcomes associated with repeat low-level MOB exposure. Outcomes of importance that are considered in this review include functional status, physical and emotional symptoms, neuropsychological outcomes, theoretically plausible biomarkers, and clinical injuries and illnesses. Human studies to date largely address samples exposed to blast during combat, making it difficult to parse primary blast exposure effects of repeated low-level MOB (i.e., injury due to the blast overpressure wave) from secondary and tertiary blast injury effects (i.e., penetrating injury from blast fragments and blunt force injury when thrown by a blast, respectively). These studies were observational, with blast exposure assessed by self-report often months or years later. Animal studies offer the important advantages of experimental design for causal inference and the capacity to isolate some effects of blast overpressure exposure, so it was decided to include animal studies that evaluated primary blast exposures of 20 psi or less. The research team was aware of concerns regarding interspecies variation in blast effects, challenges correlating level of blast in animals to those in humans, limitations of evidence gleaned from laboratory paradigms designed to model concussive rather than sub-concussive blast exposures in animals (i.e., "shock tube" methods), interlaboratory variation in procedures (e.g., position, single versus multiple exposures) and measurements, and variation as to whether exposure paradigms include multiple low-level blasts or a single blast exposure. However, to address questions of biological plausibility, it was decided to include appropriate experimental animal evidence.

A three-step process was used to develop search terms. First, potential search terms were identified from previous DoD Blast Injury Research State-of-the-Science (SoS) Literature Reviews, terms specifically relevant to the 7th SoS topic, and related National Library of Medicine Medical Subject Headings. Second, a preliminary literature search was performed and the results used to improve the search strategy. Third, the SoS Planning Committee, a group of experts from fields related to the 7th SoS topic, reviewed the search terms and recommended additional terms and search modifications. We then used a five-step process to produce this literature review: (1) definition of key questions, (2) literature search, (3) title and abstract and full text screening, (4) data abstraction, and (5) analysis. We searched peer-reviewed and gray literature that described the nature and effect of routinely incurring low-levels of MOB, including peer-reviewed scientific literature dating from 2007 to 2017, in PubMed, Web of Science, and PsycINFO, and research reports and proposals on the Defense Technical Information Center (DTIC). Additional relevant publications were identified from bibliographies of identified articles, targeted searches, and planning committee recommendations. Eligible studies included related human and animal studies and bioengineering models. This search yielded 369 full-text articles, of which 74 met inclusion criteria and were included in this review. Of these 74 articles, 28 were animal studies, 25 were human studies, and the remainder were reviews, bioengineering simulations, or medical news reports.

Of the 53 human and animal studies, 23 were longitudinal (i.e., followed subjects for 24 hours or longer). Only six followed subjects for longer than three months (three animal studies and three human studies). There was wide variation within both animal and human studies in subjects sampled, the assessments and study methods employed, and the outcomes assessed, making it difficult to reliably identify any replicated studies.

# Key Findings

## Occurrence of Low-Level MOB Exposure

The research team identified no research on the overall military population frequency of low-level MOB exposure. Five published studies evaluated military breachers—personnel who use explosives in the operational or training environment to gain rapid entry into buildings or across hard walls or structures after repeated low-level blast exposure. None of those studies combined appropriate controls, procedures to account for potential confounding, or a significant positive relationship of low-level MOB exposure to an outcome. One study found microhemorrhages in the brains and meninges of rats and pigs after experimental exposure to shoulder-mounted artillery weapon– and howitzer launcher–generated blast overpressure. There are no identified guidelines or models from which to determine what constitutes a safe level of repeated MOB exposure for the low levels often incurred in training or analogous settings.

## Potential Neurologic Effects of Low-Level MOB

### Motor Effects

None of the human or animal studies identified specifically looked for or identified motor effects of repeated low-level MOB exposure.

### Neurosensory Effects

None of the human studies specifically sought or revealed neurosensory effects of repeated low-level MOB exposure. One study of rats exposed to low-level blast found increased expression of the pain mediator transient receptor potential vanilloid 1 in corneal tissue.

### Cognition Effects

We identified two human studies of repeated low-level MOB exposure, neither of which found cognitive effects. However, six animal studies looked for cognitive effects of low-level MOB exposure; all of those showed a positive relationship, which suggests the cognitive domain may be particularly sensitive to low-level blast exposure. In these animal studies, blast exposures ranging from three to ten psi were associated with reduced learning or cognition, with measurable effects persisting up to 30 days.

### Neuropathologic Effects

None of the human studies addressed—based on our definition—specific neuropathology related to repeated low-level MOB exposure. Published studies have focused exclusively on combat samples with higher-level blast exposures. Ten of 11 studies in rodents (rats and mice) found evidence of neuropathological changes after low-level blast exposure. Findings from these studies include evidence of increased permeability of the blood-brain barrier on magnetic resonance imaging; increases in fractional anisotropy; decrease in radial diffusivity on diffusion tensor imaging; changes to the cortex and hippocampus; white-matter changes, including greater amyloid precursor protein immunoreactive cells; chronic microvascular changes; scattered pyknotic neurons; altered gene expression; dynamic microglial and macrophage responses; microdomains of brain microvessel dysfunction; and other findings.

Given the substantial interspecies neuroanatomical and cranial differences between rodents and humans, the extent to which such findings in rodents may be generalized to humans is unclear. However, the findings provide evidence that neuropathology in humans related to repeated low-level MOB exposure is plausible. Future animal studies should assess for similar effects in large animals, including nonhuman primates.

### Behavioral and Emotional Effects

Human studies, mainly cross-sectional, have suggested that behavioral and emotional symptoms following blast-related traumatic brain injury (TBI) may be largely explained by coexisting posttraumatic stress disorder (PTSD) and depression. Cross-sectional studies suggest that blast exposure may increase PTSD arousal symptoms (e.g., hypervigilance). The interplay between PTSD and TBI is complex. Blast exposure in anesthetized animals is associated with PTSD-like manifestations, leading some researchers to hypothesize that the primary injury is not psychological but instead due to direct blast exposure effects, resulting in reduced frontal lobe inhibition of the amygdala, a center of fear expression previously implicated in PTSD and the psychological threat response.

### Auditory and Vestibular Effects

It has long been known that blast exposures during military service can lead to sensorineural hearing loss. Three human studies looked for an association between blast exposure and auditory or vestibular impairments and symptoms in combat-deployed service members. Not surprisingly, all were positive. One study of 573 previously deployed service members with mTBI found a dose-response link between member-reported blast exposures and member-reported hearing loss and tinnitus. A crossover study found that low-level blast exposure from small-caliber arms fire, even while wearing fitted earplugs (adherence to earplug use was not described), affected middle-ear function and was associated with transient tinnitus. No related animal studies were identified.

### Visual Effects

Closed-globe eye injuries are known to occur from combat-related blast exposure, although these injuries typically occur at higher blast-exposure levels than those that are the focus of this review. No human studies assessing or revealing eye effects of low-level MOB exposure were identified. A single animal study assessed the eye effects of single and repeated low-level blast exposures and found increased pain and inflammation in corneal tissue.

## Early Exposure Indicators

Studies in animals and humans have examined a variety of different biomarkers with inconsistent findings to date. Preliminary studies of biosensors to monitor troops for the concussive effects of blast exposure have so far proven disappointing, and we found no studies that used biosensor data to assess subconcussive blast.

## Potential Prevention Methods

Several preventive methods are addressed, including barrier and non-barrier methods and safety guidelines. Research on barrier methods has included helmets, earplugs, and body armor. Helmets (e.g., Advanced Combat Helmet), are typically designed to protect the wearer from head trauma due to projectiles rather than blast exposure and may sometimes even amplify blast exposure. Earplugs are the most effective barrier protection from hearing-related blast exposure injury. The limited research on non-barrier prevention methods suggest that education programs may increase the use of hearing protection.

## Conclusions

### What Is Known About the Occurrence of Repeated Occupational Blast Exposure Incurred During Military Service?

We found no published information regarding military service–specific frequencies of exposure to low-level MOB. The only available information pertains to higher levels of blast exposure that is encountered in combat settings.

### What Is the Scientific Evidence Relating to the Potential Neurological Health Effects of Repeated Occupational Blast Exposure?

Experimental studies in animals suggest that persistent neurological effects from low-level blast exposure (i.e., under 10 psi) are plausible. However, interspecies differences in exposure susceptibility may be large, and there have been no experimental studies of low-level blast exposure effects in nonhuman primates. There remains significant uncertainty as to how low-level blast-exposure effects observed in animal studies translate to humans.

Epidemiological and clinical studies of military personnel provide sufficient evidence of an association between combat-related blast exposure without penetrating injury, postconcussive syndrome (PCS), and PTSD. However, these blast-exposure levels are higher than the subconcussive exposures of interest in this review, which are blast exposures of a level occurring during routine field training, breaching, artillery fire, and shoulder-mounted weapons discharge. Moreover, the precise nature of the relationship between PCS and PTSD remains unclear—it is possible that nonspecific symptoms of PTSD may explain apparent associations between mTBI and PCS. Similarly, mTBI may have effects on the amygdala that mimic PTSD-like symptoms or result in increased vulnerability to PTSD. However, the relevance of this discussion to repeated low-level MOB exposure is not known.

### What Are Promising Strategies for Preventing the Potential Neurological Effects of Repeated Military Occupational Blast Exposure?

Prevention programs targeting health risks that do not exist or implementing preventive methods that are not effective are clearly an unnecessary waste of societal resources—resources that presumably can be put to more productive use. Therefore, the relevance of discussion regarding prevention strategies depends on the answers to the following key questions that, at present, remain largely unanswered:

- *Is repeated low-level MOB exposure a significant risk to current and future force health?* There should be consensus, ideally based on empirical data, that the threat of MOB to health is significant before resources are devoted to preventing the health effects of such exposure.
- *Are current preventive interventions safe and effective?* Even if the problem is substantial, ineffective primary prevention approaches will prove wasteful.

- *Will preventive intervention benefits outweigh the harms?* If a preventive intervention is effective but renders the population vulnerable to more-serious threats, then its implementation would be self-defeating.
- *Is the preventive intervention timely and feasible?* If the preventive intervention is perfectly effective but cannot be delivered in time, it is not useful. There are any number of related factors to consider here: the availability of relevant material and staffing, and the acceptability of the intervention for leaders, service members, and society at large.

Given the early state of research into repeated low-level MOB exposure as a problem distinct from blast-related TBI, we recommend a cautious approach that would complete research into the effects of preventive intervention development and testing before implementing aggressive surveillance and prevention programs specifically targeting low-level MOB exposure.

We are not suggesting the abandonment of current protective measures against high-intensity combat blast injuries (e.g., mild, moderate, and severe TBI). However, as it pertains to low-level MOB exposure, the state of the science is preliminary at best. Implementing aggressive preventive programs against this threat without adequate evidence of preventable injury risks the consumption of considerable resources without commensurate benefit and may have adverse, unanticipated, and unintended consequences.

### What Are Promising Early Detection Indicators for the Potential Neurological Consequences of Repeated Military Occupational Blast Exposure?

The research team was unable to identify early detection biomarkers, a key type of early detection indicator, in humans. Even candidate biomarkers remain highly speculative and less than feasible, as they are almost exclusively the product of rat and mouse studies. Biosensors are a second key indicator, but we were similarly unable to identify published biosensor studies designed to assess the health effects of low-level MOB exposure. The development and validation of improved human biosensor systems and methods of using human biosensor data to model the physiologic and physical effects of low-level MOB exposure on human tissue should be prioritized and pursued. Biosensor data is a potentially low-burden method for modeling low-level MOB exposure for use in prospective cohort research designs.

## Recommendations

Perhaps the most-striking finding from this review of the literature is how little research has been done to determine the organizational threat and service member health impact of low-level MOB exposure, in contrast with our rapidly improving empirical research base relating to blast-related traumatic (concussive) brain injury. The 2018 National Defense Authorization Act mandated the design, initiation, and completion of a prospective longitudinal cohort study of low-level MOB exposure in a population-based sample of service members, and this study is one opportunity to improve understanding of repeated low-level MOB exposure. Our recommendations center on the need to improve this understanding through a coordinated program of epidemiologic, etiologic, measurement, and preventive intervention research. Organizationwide efforts to implement population exposure surveillance are more likely to be ineffective if implemented before the neurological effects of low-level MOB exposure are better characterized.

# Acknowledgments

We gratefully acknowledge Michael Leggieri, Raj Gupta, and Colonel Sidney Hinds of the Blast Injury Research Program Project Coordinating Office (PCO) for their comments, guidance, and support of this project. We also wish to recognize the extensive work that PCO had done prior to our involvement, refining the State-of-the-Science Meeting process from which this literature review has emerged, a process they have used to develop U.S. Department of Defense research policy and priorities related to blast injury since 2009.

We would also like to thank the stakeholders consulted on topic selection and the planning committee who provided invaluable guidance preparing for the 7th DoD State-of-the-Science Meeting and for this literature review.

At RAND, we wish to thank Jody Larkin and Elizabeth Hammes for their help with the literature search, Shanthi Nataraj for her thoughtful and timely consultation and support, and Kristin Sereyko for her hard work and project assistance.

We appreciate the valuable insights we received from our technical reviewers: Melinda Moore of RAND, and Gregory Elder of the James J. Peters VA Medical Center. We addressed their constructive critiques as part of RAND's rigorous quality assurance process to improve the quality of this report.

# Abbreviations

| | |
|---|---|
| ACH | Advanced Combat Helmet |
| ANAM | Automated Neuropsychological Assessment Metrics |
| AoC | alteration of consciousness |
| BBB | blood-brain barrier |
| CGRP | calcitonin gene-related peptide |
| CNS | central nervous system |
| CSF | cerebrospinal fluid |
| DoD | U.S. Department of Defense |
| DTI | Diffusion Tensor Imaging |
| DTIC | Defense Technical Information Center |
| ET-1 | endothelin-1 |
| FA | fractional anisotropy |
| LoC | loss of consciousness |
| MOB | Military Occupational Blast |
| mTBI | Mild Traumatic Brain Injury |
| OEF | Operation Enduring Freedom |
| OIF | Operation Iraqi Freedom |
| OND | Operation New Dawn |
| PCO | Blast Injury Research Program Coordinating Office |
| PCS | postconcussive syndrome |
| PTSD | posttraumatic stress disorder |
| SoS | State of the Science |
| SP | Substance P |
| TBI | traumatic brain injury |
| TRPV1 | transient receptor potential vanilloid 1 |

# CHAPTER ONE

# Introduction

Over the past decade, there has been increasing awareness of the central nervous system (CNS) effects of exposure to explosive blast. A key driver of that awareness has been the blast-related injuries suffered during combat operations in Iraq and Afghanistan (Mac Donald et al., 2011; Miller, 2012). With the relative cessation of U.S. combat operations in these regions, concern has grown over common, repetitive forms of blast exposure during military service that are, most often, unrelated to combat. Examples of these exposures include heavy weapons, such as firing artillery, recoilless rifles, and shoulder-launched rocket launchers (e.g., the Carl Gustav antitank weapon) (Hamilton, 2017). These blast exposures are of much lower intensity than those causing recognized combat-related injuries; however, becoming proficient with these heavy weapon systems may require repeated exposure. Repeated exposure scenarios raise new questions concerning the potential for effects on CNS structure, function, and subsequent development, as well as on the broader health of the service members exposed.

As awareness of the potential effects of exposure to explosive blast increases, there has been extensive study and ongoing discussion of the neurological, neurocognitive, and emotional consequences of mild traumatic brain injury (mTBI) (Hoge et al., 2008; Walker et al., 2017), but the blast exposures of interest in this report are different in both magnitude and type. At the time of exposure, mTBI is marked either by brief alterations in consciousness (AoC) (seeing stars, having one's "bell rung") or a loss of consciousness (LoC), with the latter lasting up to 30 minutes (Management of Concussion-mild Traumatic Brain Injury Working Group, 2016). In contrast, the low-level blast exposures that are of interest in this report do not result in LoC and seldom cause AoC. Furthermore, most research on mTBI to date has focused on injury due to blunt-force trauma (i.e., a physical blow to the head) rather than to blast exposure, and research demonstrates that blast-induced brain injury may be different from blunt-force trauma (Fischer et al., 2014).

The literature review that follows has therefore targeted the scientific literature pertaining to the effects of repeated, military occupational blast exposures, specifically exploring what is known regarding the nervous system consequences of these exposures. For the purpose of this review, *military occupational blast* (MOB) is defined as low-level blasts that military members are exposed to through occupational situations that do not result in immediate LoC and may not involve acute AoC or apparent symptoms. Exposure to MOB has the potential to have health or safety implications.

## Purpose of the Review

The purpose of this literature review is to support the U.S. Department of Defense (DoD) Blast Injury Research Program Coordinating Office (PCO)–sponsored Seventh DoD State-of-the-Science (SoS) Meeting. The goal of the SoS and its associated processes is to identify what is known and not known pertaining to key blast injury–related topics and emerging issues. The topic of the 7th SoS as well as this supporting literature review is "The Neurological Effects of Repeated Exposure to Military Occupational Levels of Blast: Implications for Health and Prevention."

To inform the 7th SoS, the PCO requested a literature review regarding the broad neurological and neurocognitive effects of repeated exposure to military occupational levels of blast and their implications for health and prevention. This review focuses on scientific evidence from medical, physiological, bioengineering, and health policy studies. In this review, our research team addressed the following questions:

1. What is known about the occurrence of repeated occupational blast exposure incurred during military service?
2. What is the scientific evidence relating to the potential neurological health effects of repeated occupational blast exposure?
3. What are promising strategies for preventing the potential neurological effects of repeated MOB exposure?
4. What are promising early detection indicators for the potential neurological consequences of repeated MOB exposure?

In reviewing the literature, we also sought to prioritize key research and policy gaps related to repeated MOB exposure, and examine the projects and initiatives that attempt to address them.

## Blast Exposures of Interest

Blast exposures can cause injury via primary, secondary, tertiary, quaternary, or quinary mechanisms. Primary blast exposure injuries involve tissue damage that occurs directly from the shock of the overpressure wave colliding with the body. Secondary blast exposure injuries are those produced by fragments from the exploding device, or secondary projectiles from the environment (e.g., debris, vehicle fragments). Tertiary blast exposure injuries result from blast-related displacement of body parts that strike other objects, causing a variety of injury types (e.g., blunt, avulsion, crush). Quaternary and quinary blast exposure injuries result from other explosive products or the clinical consequences of environmental contaminants (e.g., biologicals, radiation, released fuels), respectively. The magnitude of the blast exposures of interest in this literature review are exposures that do not result in LoC or AoC.

## Outcomes of Interest

We sought to review studies to help us better understand outcomes in the short (one day to three months following exposure), intermediate (three to nine months following exposure), and long terms (ten months or more after exposure). Outcomes of importance considered here included functional status, physical and emotional symptoms, neuropsychological outcomes, theoretically plausible biomarkers, and clinical injuries and illnesses.

# CHAPTER TWO

# Methodology

We used a five-step process to conduct this literature review: (1) definition of key questions, (2) literature search, (3) title abstract and full text screening, (4) data abstraction, and (5) analysis.

## Key Questions

The central review questions were: What are the neurological effects of repeated exposure to military occupational levels of blast? What are the implications for health and prevention? From these questions, our research team established the following four subquestions:

1. What is known about the occurrence of repeated occupational blast exposure incurred during military service?
2. What is the scientific evidence relating to the potential neurological health effects of repeated occupational blast exposure?
3. What are promising strategies for preventing the potential neurological effects of repeated military occupational blast exposure?
4. What are promising early detection indicators for the potential neurological consequences of repeated military occupational blast exposure?

## Literature Search

We searched peer-reviewed and gray literature that described the nature and effect of routinely incurring low levels of blast exposures in training and in theater. We searched the peer-reviewed scientific literature on PubMed, Web of Science, and PsycINFO, and the DoD grey literature on Defense Technical Information Center (DTIC) dating back to 2007. This ten-year time horizon was selected for consistency with previous SoS literature-review search criteria.

A three-step process was used to develop search terms. First, with the help of a RAND Corporation Knowledge Services librarian, potential search terms were identified from previous SoS reports, terms specifically relevant to the 7th SoS topic, and related Medline medical subject headings. Second, a preliminary literature search was performed and its results used to improve the initial search strategy. Third, the SoS planning committee reviewed the search terms and recommended additional terms. Search terms are in Table 2.1.

**Table 2.1**
**Search Terms**

| | Search Terms |
|---|---|
| Exposure | Blast; brain; combat, battle; training, field training; low-level; psi; pressure, overpressure; shoulder-fired weapon; training; trauma; artillery, gun, cannon, rifle, howitzer, mortar, recoil, grenade, bomb, bazooka; Carl Gustaf / Carl Gustav |
| Population and context | Breachers; military, Army, Navy, Marine; occupational; deploy, deployed, deploying; infantry; Occupational Safety and Health Administration (OSHA); mining; field exercise, military exercise |
| Strategies, interventions, and challenges | Armor; helmet; Kevlar; goggles; detection; mitigation; prevention; protection, protective gear; mitigation; safety–field training, military, workplace, occupational; combat simulation, battle simulation |
| Injury or outcomes | Concuss*; disability; functional status; memory; central nervous system; ocular; auditory; headache; neuro*; lung; cerebrospinal; mental health; traumatic brain injury, mTBI; biomarker; seizure; epilepsy; posttraumatic stress disorder (PTSD); cognitive deficits; axonal injury; white matter; EEG (electroencephalogram); working memory; neuroendocrine; neuroinflammation; neurovascular; neurocognitive; neuronal |

# CHAPTER THREE

# Review Findings

Our search yielded 3,892 citations for initial title and abstract screening. Two teams of two reviewers each screened article abstracts for human or animal research examining low-level blast exposure and its relationship to neurologic health, including hearing and vision. To achieve reliability, reviewer teams compared results, and then discussed any discordant codes between themselves and with the review team and principal investigator.

Title and abstract screening yielded 255 citations. Reference lists were screened and the SoS planning committee consulted. Targeted research team searches yielded an additional 114 full text articles (studies, clinical and safety guidelines, policy documents, other relevant grey literature, past DoD Blast Injury Research SoS Meeting literature reviews and reports). The resulting 369 full-text articles were carefully read, and prespecified inclusion and exclusion criteria were applied (see Table 3.1 and Figure 3.1). Full-text review and abstraction was completed to categorize articles by their type, target population, research design, and length of follow-up (if longitudinal information was provided). We excluded (1) human or animal studies that did not provide information regarding a blast exposure injury or a blast exposure assessment; (2) human studies that evaluated only severe blast exposure injuries (e.g., moderate or severe traumatic brain injury (TBI); mTBI but with no comparisons that excluded individuals with LoC); (3) animal studies with no experimental group receiving less than 20 psi blast exposure (see following section, Defining Low Level Blast Exposure); or (4) studies focused on pathologies outside of the scope of the review (e.g., psychiatric conditions in the absence of blast exposure injury assessments). After full text review, 74 articles met these final inclusion and exclusion criteria. Of these 74 studies, 28 were animal studies and 25 were human studies. The remainder were reviews ($n = 7$), bioengineering simulations ($n = 10$) or medical news reports ($n = 4$).

**Table 3.1**
**Literature Review Search Criteria**

| Inclusion Criteria | Exclusion Criteria |
|---|---|
| English language articles only | Articles not directly addressing research questions |
| Articles published between 2007 and 2017 (inclusive)[a] | DTIC documents not approved for public release |
| Clinical and animal model studies | |
| DTIC documents assigned Distribution A: Approved for public release: distribution unlimited | |

[a] Older publications were included when they were potentially critical to addressing the research questions or understanding the topic.

**Figure 3.1**
**Literature Search Flow Diagram**

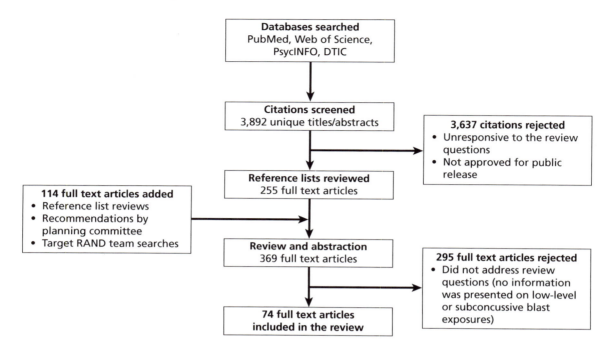

Of the 53 human and animal studies, 23 were longitudinal. Of the longitudinal studies, 16 studied animals (Baalman et al., 2013; Elder, Dorr, et al., 2012; Ewert et al., 2012; Gama Sosa, De Gasperi, Paulino, et al., 2013; Heldt et al., 2014; Kamnaksh et al., 2014; Luo et al., 2014; Park et al., 2011; Perez-Garcia, Gama Sosa, et al., 2016; Pun et al., 2011; Rodriguez et al., 2016; Rubovitch et al., 2011; Säljö, Bolouri, et al., 2010; Säljö, Mayorga, et al., 2011; Tweedie et al., 2013; Woods et al., 2013) and seven studied humans (Carr, Stone, et al., 2016; Haran et al., 2013; Parish et al., 2009; Shupak et al., 1993; Tate et al., 2013; Thiel, Dretsch, and Ahroon, 2015; Walker et al., 2017). Only six studies (Elder, Dorr, et al., 2012; Gama Sosa, De Gasperi, et al., 2014; Haran et al., 2013; Luo et al., 2014; Thiel, Dretsch, and Ahroon, 2015; Walker et al., 2017) followed subjects for longer than three months (three animal studies, three human studies) and only three followed subjects for longer than nine months, all of which were human studies (Haran et al., 2013, Thiel, Dretsch, and Ahroon, 2015; Walker et al., 2017). Among the longitudinal studies, both animal studies and human studies were noted to vary widely by sample, exposure methods, and assessments, as well as by outcome domains and assessments. As a result, it was difficult to clearly identify any replicated findings.

## Defining Low-Level Blast Exposure

A full discussion of the biophysics of blast exposure is beyond the scope of this review. The interested reader can find a frequently cited and informative introduction in the Institute of Medicine report, *Gulf War and Health*, Vol. 7: *Long-Term Consequences of Traumatic Brain Injury* (Committee on Gulf War and Health, 2008). The direct physical effects of blast exposure are complex and depend on the intensity and duration of the blast, distance of an exposed object from the originating blast, and the presence of reflective barriers (e.g., nearby hard surfaces or enclosures) that may serve to amplify the exposure effects of a single blast event. There are important interspecies differences in the brain effects of blast exposure (Elder, Stone, and Ahlers, 2014; Säljö, Bolouri, et al., 2010). In humans, a blast exposure of 100 psi is widely considered lethal, and between 60 and 80 psi is potentially fatal. Anterograde memory deficits without neurological impairment or gross neuropathological or neurohistological changes are produced in rodents after a 5.3 to 10.8 psi shock-tube blast exposure (Säljö, Bolouri, et al., 2010). This finding is consistent with other studies suggesting that, in rats, blast exposures of up to 10.8 psi are consistent with low-level blast exposure (Ahlers et al., 2012). In contrast, blast exposures of 17 psi produced subdural hemorrhages and cortical contusions in rats (Ahlers et al., 2012), a level of neuropathology more severe that is found in neuroimaging studies of humans with mTBI. A summary of this information is shown in Table 3.2.

Therefore, in an effort to be maximally inclusive of available research evidence relevant to low-level MOB exposure, we included animal studies that assessed the effects of blast exposures up to 20 psi. It is important to note, however, that there is substantial variation across studies with regard to exposure paradigms, exposure thresholds, other blast characteristics (e.g., multiple versus single), animal species, pressure sensors, position of the animal at the time of exposure, and other laboratory-related factors that can affect blast-related exposure outcomes. However, given the inability to use experimental research designs in humans and the small number of human studies addressing low-level MOB exposures, the research team opted to include appropriate experimental animal evidence to inform questions of biological plausibility.

## Magnitude of the Problem

The research team found no generalizable military-wide or service-specific population data (or ongoing studies) from which to estimate the occurrence of repeated, low-level MOB exposure or its potential health consequences.

**Table 3.2**
**Sample Levels of psi Experienced in Training and Theater**

| Level of Blast | Impact on Animals or Humans |
| --- | --- |
| 3.3 psi | Sustained intracranial pressure increases in pigs exposed to firing of military weapons |
| 4.4 psi | Sustained intracranial pressure increases detected in rats |
| 5.1 psi | Positional transmission to rat brain |
| 5.3–10.8 psi | Resulted in anterograde memory deficits without neuropathological or histological changes in rats |
| 17 psi | Produced subdural hemorrhages and cortical contusions in rats |
| 60 – 80 psi | Potentially fatal blast exposure in humans |
| 100 psi | Lethal blast exposure in humans |

SOURCES: Ahlers et al., 2012; Budde et al., 2013; Chavko et al., 2011; Säljö, Arrhén, et al., 2008; Säljö, Bolouri, et al., 2010; Säljö, Mayorga, et al., 2011.

## Incidence and Prevalence of Blast Exposure

According to the Defense Veterans Brain Injury Center, blast exposure injuries have been responsible for over 65 percent of the casualties in the ongoing conflicts in Iraq and Afghanistan (Wojcik et al., 2010). However, injury predication models (e.g., the Bowen survivability curve) used by DoD are limited to blasts more powerful than those incurred during training. Consequently, there has been little research and few (if any) guidelines or models from which to determine what may constitute a safe number or level for the low levels of blast exposure often incurred in training or analogous settings (Teland, 2012). We identified no research on the overall frequency with which low-level MOB exposure occurs. However, there have been several studies of potentially high-risk populations, such as explosive breachers, shoulder mounted artillery operators, and military service members deployed and in training.

## Studies of Breachers

Military breachers are personnel who use explosives in the operational or training environment to gain rapid entry into buildings or across hard walls or structures. Breachers are regularly exposed to subconcussive levels of blast. There have been anecdotal reports of breachers who have experienced cognitive problems and other symptoms after extensive exposure to low-level MOB. We identified five studies that investigated this issue, none of which combined appropriate controls, attempted accounting for potential confounding, and a significant positive relationship of low-level MOB exposure to an outcome. A conference abstract described a pre-post study of 31 Marine breachers without a control group for comparison. After two weeks of follow-up, most cognitive indicators showed improvements from the pre-exposure baseline, which the authors attributed to probable learning effects (Parish et al., 2009). There were no significant declines in cognitive function. Thiel and colleagues (2015) studied 12 Marine breacher trainers and compared them with a control group of 28 unexposed breacher engineers. Ten of the breachers were followed for two years—the researchers found no association between cumulative subconcussive blast exposure from routine breacher training results and persistent neurological manifestations (Thiel, Dretsch, and Ahroon, 2015).

In another study, a surveys of symptoms was administered to 135 breachers and 49 non-breachers, finding significantly higher numbers and severity of self-report symptoms among the breachers. However, results were not adjusted for important and potentially confounding variables, such as TBI history (Carr, Polejaeva, et al. 2015; Carr, Stone, et al., 2016). Yet another study compared five breacher instructors with 26 breacher students after breacher training and seven Marine controls on a broad battery of standard neurocognitive tests: the Automated Neuropsychological Assessment Metrics (ANAM) TBI battery, self-report measures (e.g., blast exposure, anxiety, depression, PTSD), and functional magnetic resonance imaging. The results did not yield clear evidence of neurological impairment in breachers or instructors (Carr, Stone, et al., 2016). Another study of 21 New Zealand breacher trainees attempted to link a number of experimental serum biomarkers of possible TBI to neurocognitive performance and self-reported symptoms before, during, and after a two-week breacher training course (Tate et al., 2013). Some biomarkers were indeed associated with symptoms, but there was no direct measurement of exposure provided from which we could link to either biomarkers or symptoms, and no control group of unexposed individuals for comparison (Tate et al., 2013). In sum, the breacher studies to date have involved small samples, resulting in limited power to detect potentially important differences and in limitations to generalizability. Carefully designed large, longitudinal studies that assess potential confounders, achieve high follow-up rates, and use validated outcome measures are needed to better understand and describe the neurocognitive, functional, and symptom outcomes associated with this potentially high-risk occupation.

### Studies of Shoulder-Mounted Artillery Operators

Shoulder-mounted weapons have become an increasingly powerful and important combat weapon. The blast generated from the Carl Gustav and shoulder-launched multipurpose assault weapon (SMAW) has not been quantified (Wiri et al., 2017). A single round from the Carl Gustav can weigh nearly ten pounds and is powerful enough to destroy a tank (Hamilton, 2017).

We found one study in this category. Investigators at the Swedish Sahlgrenska Academy found microhemorrhages in the brains and meninges of rats and pigs after exposure to experimental Carl Gustav– and howitzer launcher–generated blast overpressure. Subsequently, the Swedish Armed Forces are said to have restricted the number of rounds that service members can be exposed to daily (Säljö, Arrhén, et al., 2008; "Blast Overpressure Is Generated from the Firing of Weapons, and May Cause Brain Injury," 2009). However, we found neither experimental animal studies replicating these findings nor descriptive or analytic epidemiologic studies to suggest human effects of these or similar weapons systems, and the current status of the Swedish Armed Forces policy is uncertain.

### Combat Settings

Combat-related health issues have been described expansively in the literature. However, our research team found no data regarding repeated combat exposures to low-level MOB. Studies have shown that combat exposure to concussive blast is associated with sensory symptoms, including complaints of hearing loss and tinnitus (Reid et al., 2014) and poorer cognitive performance (Haran et al., 2013). However, we did not identify any epidemiologic studies that specifically evaluated the health effects of low-level MOB exposures during combat.

### Other Occupations or Informative Analogues

The research team expanded its search from military and biomedical resources to include occupational safety, labor, physics, and engineering literature. We sought to learn from the mining and construction industries, which interact with small blast ordnances on a routine basis. While there are likely many parallels, the literature reviewed in this section emphasized pulmonary, dermatologic, and other morbidities associated with inhalation of toxic particulates, instead of the cumulative neurological effects of blast exposure.

## Potential Neurological Consequences and Mechanisms

### Motor Effects

None of the human or animal studies identified specifically looked for or identified motor effects of repeated low-level MOB exposure.

### Neurosensory Effects

None of the human studies specifically sought or revealed neurosensory effects of repeated low-level MOB exposure. One study of rat corneal tissue found that exposure to both single and repeated low-level blast was associated with increased transient receptor potential vanilloid 1 (TRPV1) expression (Por, Choi, and Lund, 2016). TRPV1 has a role in mediating various types of pain, and it is expressed on small, unmyelinated sensory neurons in the trigeminal ganglia.

## Cognition Effects

Two human studies evaluated neurocognitive effects of repeated low-level MOB exposure, both negative studies of breachers (Parish et al., 2009; Tate et al., 2013). Two other studies looked at cognitive effects in samples that contained service members with mixed concussive and subconcussive blast exposure. Haran and colleagues (2013) analyzed longitudinal ANAM test data from 169 recently deployed, highly combat-exposed marines. Many reported concussions, and concussion was associated with measurable decreases in cognitive performance 2–8 weeks after deployment. Following up with the subjects after eight months, postconcussive symptoms persisted, but measurable cognitive deficits had resolved (Haran et al., 2013). A review of the cognitive construct assessments from nine studies (12 groups) involving 1,154 participants with more severe repeated combat blast–related mTBI—50 percent reported associated LoC (Karr, Areshenkoff, and Garcia-Barrera, 2014)— found that executive function, verbal delayed memory, and processing speed were the most-sensitive cognitive domains to blast-related TBI, and that observed associations were not explained by PTSD symptom severity. All of the studies in this review were cross-sectional and received low quality ratings. The average participant was assessed an average of 3.8 years after the event (a factor plaguing the vast majority of military and veteran studies of mTBI from any cause to date). The findings, however, suggest that cognitive effects of low-level, repeated MOB exposure are plausible and may not be explainable by concurrent PTSD.

We identified six animal studies that investigated low-level blast exposure, all of which revealed positive findings. Studies exposing rats to repeated low-level blast have found anterograde memory deficits on a passive avoidance task at 10.8 psi, and transient learning deficits in a water-maze task after 5.3 psi blast exposures (Ahlers et al., 2012); memory deficits in a water maze persisting through to a 30-day follow-up after exposure to a single 14.5 psi blast (Budde et al., 2013); and impaired memory on a novel-object recognition task after a single exposure to 10.7 psi (Baalman et al., 2013). A study of mice found impaired object recognition after multiple 2-psi blast exposures paired with 100 decibel noise (Xie, Kuang, and Tsien, 2013); impaired memory on novel-object recognition and Y-maze for 30 days after a single 5.5-psi blast exposure (Rubovitch et al., 2011); and impaired memory on novel-object recognition but not Y-maze or passive avoidance at 30 days after a single 2.5-psi blast exposure (Tweedie et al., 2013). Though it is unclear how such findings may translate to humans, these studies suggest that it is plausible that there would be cognitive effects after repeated, low-level MOB exposure. Studies in large animals, with a longer period of follow-up, are needed.

## Neuropathology Effects

None of the 25 human studies we reviewed looked for neuropathologic changes after low-level MOB exposure. However, eleven animal studies investigated neuropathological changes after exposure to low-level blast exposure, and all but one (Elder, Stone, and Ahlers, 2014) found changes. An anatomy-based region-of-interest analysis using diffusion tensor imaging (DTI) to study rat brains (Kamnaksh et al., 2014) found significant interactions in axial and radial diffusivity in a number of subcortical structures after single and multiple 19.9-psi blast exposures. Multiple (versus single) exposed rats were associated with thalamic (but not hippocampal) changes. In rats two weeks after a single 10.7-psi blast exposure, Baalman and colleagues (2013) found little or no changes in cerebral cortex, corpus callosum, and hippocampus injury markers, though in the blast-exposed animals there was significant shortening of axon initial segments in the hippocampus and cortex. Another study found evidence of white-matter damage and minimal cell death, localized mainly in the corpus callosum and periventricular regions in rats after a 1.7-psi blast exposure (Park et al., 2011). There was also evidence of shear lesions in rats, and chronic changes in the microvasculature were evident several months after exposure to both single and multiple 10.8-psi blasts (Gama Sosa, De Gasperi, et al., 2014). Another study found an increase of intracranial pressure several hours after low-level (4.4- or 8.7-psi) exposure to a single blast (Säljö, Bolouri, et al., 2010). Scattered pyknotic neurons have also been found in the rat cortex after exposure to two blasts of 2.9 psi (Moochhala et al., 2004; Pun et al. 2011). Altered gene expression in over 5,700 genes, as well as greater amyloid precursor protein immunoreactive cells in white matter, was also observed following a single low-level blast exposure (Pun et al., 2011). Finally, decreases in fractional anisotropy (FA) have also been observed in brains that were exposed to a single low-level blast exposure (Budde et al., 2013).

Among mouse models, one study found increased ganglioside and depleted ceramide in the hippocampus, thalamus, and hypothalamus associated with low-level blast exposure (Woods et al., 2013). Researchers also observed an increase in blood-brain barrier (BBB) permeability one-month post-single low-level blast exposure on MRI (Rubovitch et al., 2011). While DTI showed an increase in FA and a decrease in radial diffusivity, these changes may represent brain axonal and myelin abnormalities. Other studies using mouse models reported acute subcortical changes after mild blast-induced TBI, with both single and multiple blast exposures (Kamnaksh et al., 2014). Finally, a robust astrogliosis and increased p-Tau immunoreactivity was observed upon post-mortem pathological examinations of mice after low-level blast exposure (Luo et al., 2014). Interestingly, one mouse model explored the differences in blunt-force trauma mTBI and blast-related mTBI and found them to be distinct, in that pathways that lead to Alzheimer's disease are up-regulated in blunt-force-trauma injury and downregulated in blast injury (Tweedie et al., 2013). Finally, researchers found single low-level blast exposure caused dynamic microglial and macrophage responses and microdomains of brain microvessel dysfunction among mice (Huber et al., 2016). They also argue that mild blast exposure causes an evolving CNS insult that is initiated by discrete disturbances of vascular function. Both of these conditions set the stage for more-protracted and more-widespread neuroinflammatory responses.

Given substantial interspecies neuroanatomical and cranial differences between rodents and humans, the extent to which such findings in rodents may be generalized to humans is unclear. However, the findings provide evidence that neuropathology in humans, related to repeated low-level MOB exposure, is plausible. Future animal studies should assess for the neuropathological effects of low-level blast exposure in large animals, including nonhuman primates.

## Behavioral and Emotional Effects

Four human studies that we reviewed assessed behavioral and emotional issues related to mixed samples of concussive and other blast exposures. Studies of blast-related mTBI suggest that blast exposure may be associated with a myriad of symptoms, both behavioral and emotional (e.g., PTSD, depression, anxiety) and physical (including those symptoms we think of most classically as "neurological" symptoms, such as headache, dizziness, memory problems, confusion, and nausea) (Hoge et al., 2008; Wilk et al., 2012). These studies suggest that mental and physical symptoms are elevated in blast-related mTBI, though to a lesser degree, even in the absence of a precipitating head injury with LoC (Hoge et al., 2008; Wilk et al., 2012). A possibility that both of these studies raise is that postconcussive symptoms are epiphenomena of PTSD and depression. Macera et al. (2012), however, investigated the co-occurrence of mTBI and PTSD and found that blast-related TBIs worsen the self-report of symptoms that overlap with PTSD, such as irritability (Macera et al., 2012). Likewise, Toyinbo et al. (2017) conducted an an online survey and found that blast exposure, not mTBI per se, was associated with greater PTSD arousal symptoms and tinnitus (Toyinbo et al., 2017) and mTBI diagnosis alone was not significantly associated with an increase in other PTSD symptoms (Toyinbo et al., 2017). Unfortunately, however, we identified no human studies that isolated the effects of low-level MOB exposure on behavioral and emotional problems.

While it is tempting to view blast-related TBI and PTSD as distinct disorders, Elder, Stone, and Ahlers (2014) have offered a provocative model in which blast exposure may unmask a vulnerability to PTSD-related symptoms. In this model, blast-related TBI may reduce frontal cortical inhibition of the amygdala, a center of fear expression thought to heighten responses to psychological threats, which is implicated in PTSD development. For example, the same research team found that a blast-exposed rat, many months after the blast exposure, develops new traits that weren't there before the exposure when exposed to a predator scent (a psychological stressor). They argue that an initial blast exposure may predispose individuals to develop PTSD in response to a subsequent psychological stressor. This can be seen as further support for blast exposure affecting the biological substrates that underlie PTSD (Perez-Garcia, Gama Sosa, et al., 2016). In a separate study, the authors showed that these PTSD-related symptoms were often present many months after blast exposure, and that these symptoms could be reversed with the medication BCI-838, which is currently under review for use in humans for depression and suicidality, and has been shown to have antidepressant effects in animals (Perez-Garcia, De Gasperi, et al., 2018).

Studies of animals exposed to a low-level blast also offer valuable insight into the potential association of behavioral symptoms and low-level MOB exposure. We identified five animal studies that investigated this issue, all of which had positive results. Rat and mouse exposure to a single low-level blast was associated with a decreased preference for novel objects (Rubovitch et al., 2011; Baalman et al., 2013; Tweedie et al., 2013), and multiple and single exposures were associated with increased anxiety, enhanced learned-fear response, and intensified acoustic startle, though in most of these studies the animals are never aware of a blast-induced traumatic stressor because they are anesthetized (Elder, Dorr, et al., 2012; Heldt et al., 2014). This may lend support for Elder and colleagues' frontal lobe-amygdala fear-disinhibition hypothesis.

## Auditory and Vestibular Effects

It has long been known that military populations are exposed to a multitude of compounding risks for auditory impairment, potentially leading to inner-ear and cochlear injuries (e.g., blast wave and noise exposure), middle-ear injuries (e.g., tympanic membrane perforation, ossicular disruption), outer-ear injuries (e.g., burns, flying debris), and chronic tinnitus ("ringing of the ears") (Fausti et al., 2009). We identified three studies of blast exposure on the auditory and vestibular system in humans, and perhaps not surprisingly, all of them yielded positive findings. One study of 573 previously deployed service members with mTBI found a dose-response relationship between higher levels of service member–reported combat-blast exposure and service member–reported hearing loss and tinnitus (Reid et al., 2014). Using a crossover design, researchers found that exposure to repeated low-level blasts from small-caliber firearms affected middle-ear function and was associated with subsequent report of transient tinnitus, even when wearing fitted earplugs, though adherence to earplug use was not reported. They concluded that exposure to small-caliber firearms may play a role in the early stages of auditory fatigue and eventual hearing loss (Job et al., 2016). The vestibular system helps to maintain balance and postural stability and has the potential to be impacted by repeated exposure to low-level blasts, though these effects are not well understood. The third study described five Israeli soldiers exposed to blast but who did not experience head trauma, LoC, or amnesia. Three reported vertigo, four had hearing loss, four suffered from tinnitus, and one had otalgia (Shupak et al., 1993).

### Visual Effects

There were no human studies found of eye effects of low-level MOB exposure. Closed-globe eye injuries are known to occur from combat-related blast exposure, though these injuries typically occur at higher blast-exposure levels than are the focus of this review. We found one animal study on the eye effects of low-level blast exposure and it yielded positive findings. Por and colleagues (2016) found that both single and repeated 9.9-psi blast exposures led to increased pain and inflammation markers in the corneal tissue of rats.

## Potential Early Indicators of Low-Level MOB Exposure

There are significant disadvantages associated with measuring low-level MOB exposure. There are few currently available methods, and Table 3.3 describes their advantages and disadvantages. Most of these assessment methods involve a significant burden for service members or the military unit. None of the methods are supported by published assessments of validity, reliability, adherence, or feasibility for assessing low-level MOB exposure.

One promising way to detect low-level MOB exposure may be through biomarkers and sensors. An accurate, reliable, and lightweight biosensor could offer objective, real-time assessment of exposure with minimal distraction to service members and military units. We identified three biosensor studies, none of which suggested that biosensors were effective for their intended purpose. One of these was an unpublished study suggesting the biosensors were insufficiently sensitive or specific for surveillance of combat blast–related TBI (Department of Defense Blast Injury Research Program Coordinating Office, 2014). These preliminary data suggest that there may be insufficient standardization of available biosensors for assessing low-level MOB exposure.

The Defense Advanced Research Projects Agency (DARPA) funded the development of the Blast Gauge System, which is a three-piece sensor system designed to record overpressure exposure levels in the battlefield. While the sensor technology may be adequate, a major challenge in this field is the absence of a feasible way to link sensor data to blast exposure injury. Panzer et al. (2012) demonstrated that in individuals with severe non-MOB blast-induced TBI, injury tolerance decreased with each exposure. Individual differences in overpressure-exposure thresholds to injury may present difficulties in the development of a general guideline for over-

**Table 3.3.**
**Methods of Assessing Exposure to Low-Level MOB**

| Assessment Method | Advantages | Disadvantages |
| --- | --- | --- |
| Service member or buddy self-report | • Simple<br>• Low-cost | • Significant unit workload<br>• Significant service member workload<br>• Easily gamed<br>• Requires ongoing quality control |
| Supervisor report | • Accountable | • Significant leader workload<br>• Low priority under fire<br>• Low feasibility |
| Independent research rater | • Low unit workload<br>• Quality control-data gathering | • Challenge embedding in unit |
| Biomarkers | • Objective indicator<br>• Harder to game | • No agreement on best biomarker(s)<br>• May require biological samples<br>• Potentially costly sample storage and processing |
| Biosensor | • Objective indicator<br>• Low burden<br>• Acceptable to assessor | • Device reliability and validity concerns |

pressure exposure limit and frequency (Panzer et al. 2012). Second, Courtney and Courtney (2011) identified that such thresholds are subject to blast exposure conditions and that a blast exposure can induce brain injury through multiple simultaneous mechanisms. This indicates that how an exposure event occurred (blast-specific) affects injury, and that relationship may need to be quantified. Finally, McEntire et al. (2010) showed that helmet-mounted sensors do not measure the overpressure that the head experiences during a blast. Their experimental data revealed that the head may experience vastly different forces, despite similar helmet sensor readings (McEntire et al. 2010).

Work has also been done in the area of biomarkers. We identified four studies within this category, two of which had positive results. As mentioned above, Por et al. (2016) exposed rats to single, multiple, or zero (control) compressed-air blasts to determine the expression of the TRPV1 channel, calcitonin gene-related peptide (CGRP), substance P (SP), endothelin-1 (ET-1), neutrophil infiltration, and myeloperoxidase (Por et al., 2016). These were identified as associated with blast exposure and pain and inflammatory mediators following ocular trauma. Results showed an increased expression of TRPV1, CGRP, SP, ET-1, and neutrophil infiltration; therefore, these findings suggest activation of pain- and inflammation-signaling following blast exposure (Por et al., 2016). Another study by Tate and colleagues (2013) looked at biomarkers in New Zealand Defense Force members who are breachers. This study found higher concentrations of three biomarkers—ubiquitin C-terminal hydrolase-L1, aII-spectrin breakdown product, and glial fibrillary acidic protein—which were associated with significantly longer reaction times, fewer correct answers on neurocognitive performance tests, and increased symptom reporting (Tate et al., 2013). Unfortunately, the study did not directly measure low-level blast exposure or employ a control group differing in low-level blast exposure. Therefore, the study did not validate these biomarkers as indicators of low-level MOB exposure.

Blennow and colleagues (2011) completed three studies to assess the relationship of repeated low-level MOB exposures to various biomarkers. First, 21 Swedish military officers were asked to repeatedly fire a howitzer or bazooka; then the researchers obtained samples of their cerebrospinal fluid (CSF) and serum biomarkers. Second, another group of 32 officers fired high-explosive antitank grenades using Carl Gustaf–model bazookas; serum biomarkers were drawn the day before; 30 minutes after; and one, 12, and 24 hours after exposure. Third, seven officers were exposed to 100 charges of detonating explosives over ten days. Serum samples were drawn the day before exposure; days 8, 9 and 10 of the exposures; and then ten days after all exposures. Nineteen healthy, age-matched volunteers acted as comparison controls for all three exposed groups. CSF biomarkers for neuronal or axonal damage (tau and neurofilament protein), glial cell injury (glial fibrillary acidic protein and S-100 calcium-binding protein B), BBB damage (CSF and serum albumin ratio) and hemorrhages (hemoglobin and bilirubin) were unrelated to the blast exposures (Goverover and Chiaravalloti, 2014).

The magnitude of the biomarker challenge is perhaps best captured by Elder and colleagues, who point out, "there are no biomarkers that can distinguish cognitive, affective, and somatic symptoms induced by a psychological stressor from those induced by physical trauma" (Elder et al., 2014). All of these factors similarly complicate our ability to understand and analyze blast exposure sensor data. The development and use of improved sensors and biomarkers, along with in situ coordination of care, will be instrumental to addressing this knowledge gap.

## Prevention Strategies

"Prevention is the best medicine" remains a pervasive truism and a widely held and largely unchallenged societal view. However, prevention programs targeting health risks that do not exist or implementing preventive methods that are not effective is clearly an unnecessary waste of societal resources, resources that presumably can be put to more productive use. Therefore, the relevance of the following discussion (and future discussions) of prevention strategies depends on the answers to several key questions that at present remain largely unanswered:

- *Is low-level MOB exposure a significant risk to current and future force health?* There should be general consensus, ideally based on empirical data, that the threat to health posed by MOB is both real and significant before we devote significant resources to preventing the health effects of an exposure.
- *Are current preventive interventions safe and effective?* Even if the problem is substantial, ineffective primary prevention approaches will prove wasteful.
- *Will preventive intervention benefits outweigh the harms?* If a preventive intervention is effective but causes more harm than good, then implementation is likely self-defeating.
- *Is the preventive intervention timely and feasible?* If the preventive intervention is perfectly effective but cannot be delivered in time or in the appropriate context, then it is not useful. There are any number of related factors to consider here, such as the availability of relevant material and staffing and acceptance of the intervention by leaders, service members, and the larger society.

While prevention of illness and injury associated with repeated exposure to low-level blasts is relatively underresearched, we can draw from the literature on prevention methods for blast exposure injuries more generally to better understand prevention methods. The main types of prevention methods are barrier methods—for example, helmets, earplugs, and protective goggles.

## Protection Methods

Research indicates that helmets that service members wear, such as the Advanced Combat Helmet (ACH), are traditionally designed to protect wearers from head trauma caused by projectiles. Unfortunately, they are not designed to protect wearers from blast overpressure exposure and resulting neurotrauma. Moreover, traditional helmets may even exacerbate blast-related injury (Grujicic, Bell, Pandurangan, and Glomski, 2011; Ganpule et al., 2012; Kulkarni et al., 2013; Moss, King, and Blackman, 2009). While some studies found that ACH provided some level of protection from blast-related injury (Grujicic, Bell, Pandurangan, and Glomski, 2011), others found that space between pads inside the helmet creates blast-pressure-wave access to the skull and brain. When wearers are exposed to blast overpressure waves, these spaces allows the blast waves under the helmet where they reverberate, occasionally amplifying them in excess of the external blast-pressure waves. This causes the skull to flex, producing potentially dangerous forces acting on brain tissue (Moss, King, and Blackman, 2009).

One promising solution is to produce helmets without gaps between pads. In models, this has been shown to greatly reduce the reverberation of the blast within the helmet (Moss, King, and Blackman, 2009; Ganpule et al., 2012). Other potential solutions include the use of polyurea foam, face shields, and earplugs. Polyurea foam, compared with standard foam paddings, reduces the peak load of the blast wave (Grujicic, Bell, Pandurangan, and He, 2010). A partial face shield may reduce stress intensity on the brain (Nyein et al., 2010). Earplugs are the most effective barrier against blast-related hearing loss (Helling, 2004; Schulz, 2004; Wilmington et al., 2009; Dougherty et al., 2013). The angle of the head in relation to a blast exposure plays a role in how explosive force may cause the skull to flex (Chavko et al., 2011) and should be considered in the design of weapon systems that could result in repeated low-level blast exposure and how these weapons are used during training.

Body armor also has the potential to protect against chronic damage as a result of exposure to blasts. Rodriguez et al. (2016) found that mice protected with a polycarbonate body shield during blast exposure experienced a lower degree of signal enhancement compared with mice lacking a body shield, suggesting that improved body armor could reduce risks associated with blast exposure.

## Nonbarrier Methods

Nonbarrier prevention methods, such as education programs, may aid efforts to reduce low-level MOB exposure. Available studies primarily address hearing loss. Among carpenters, research has suggested that education programs can increase the usage of hearing protection methods (Stephenson and Stephenson, 2011). However, one study found that even though British soldiers knew the specifics of hearing protection policy and that their job could affect their hearing, several factors prevented proper protection usage: communication difficulties, discomfort, and impracticability of use in some situations. The investigators concluded that hearing conservation education programs should be uniquely tailored for military populations (Okpala, 2007). An experimental rat study found that the antioxidant 2,4-disulfonyl alpha-phenyl tertiary butyl nitrone (HPN-07) combined with N-acetylcysteine (NAC) administered an hour after three consecutive 14-psi blast exposures enhanced recovery and prevented permanent hearing loss (Ewert et al., 2012). While a promising lead, the work is preliminary (Oishi and Schacht, 2011).

## Ongoing Research

The military is currently funding ongoing research regarding protection from occupational blast exposure. One study is led by Rong Gan at the University of Oklahoma. The objective of this study is "to determine middle ear protective mechanisms and develop the finite element (FE) model of the human ear for simulating blast exposure injury and assisting design/ evaluation of [hearing protective devices] HPDs" (Gan, 2015). The study is entering its final year (as of this writing in 2019) and is attempting to validate HPDs using the FE model that the researchers have developed. Another ongoing project, led by Brittany Coats at the University of Utah, is studying "Temporal Progression of Visual Injury from Blast Exposure" (Coats and Shedd, 2016). This study found that when rats are exposed to shock tube blast, the stroma of the eye thickens after two weeks, then the epithelial layer thickens at five weeks, leading to eventual corneal scarring. Early identification of corneal thickening may provide a time window during which appropriate early intervention could prevent corneal scarring and resulting visual impairment.

# CHAPTER FOUR

# Discussion

This section starts with a general summary of central findings, using the four key questions outlined in Chapter One as guideposts. Then, for each question, we discuss and attempt to prioritize key research gaps and opportunities for further understanding related to low-level MOB exposure. The questions are

1. What is known about the occurrence of repeated occupational blast exposure incurred during military service?
2. What is the scientific evidence relating to potential neurological health effects of repeated occupational blast exposure?
3. What are promising strategies for preventing the potential neurological effects of repeated MOB exposure?
4. What are promising early detection indicators for the potential neurological consequences of repeated MOB exposure?

For the research questions, we did not identify any epidemiological studies of low-level MOB exposure to help us understand the potential magnitude of the issue. We explored such specific situations and assignments as breacher training, shoulder-mounted artillery operators, and military service members deployed and in training—however, none of this research offers a comprehensive understanding of the magnitude of low-level MOB and whether low-level MOB exposure constitutes a neurological risk.

## Research Opportunities and Gaps

In this section, we discuss the research necessary to develop a strong understanding of the neurological effects of low-level MOB exposure. A full understanding of the potential short- and long-term outcomes of low-level MOB is challenging because there has been little research on this topic to date, and the research that does exist does not present a clear picture of the issue. Most research on the issue of blast-related brain injury is generally concerned with a magnitude of blast exposure that is stronger than low-level MOB exposure. And, while epidemiologic and clinical studies to date provide sufficient evidence of an association between combat-related blast exposure without penetrating injury and postconcussive syndrome (PCS) and PTSD, this is not the type of blast exposure that is the focus of this report. Furthermore, the nature of the relationship between PCS and PTSD remains unclear, and it is possible that nonspecific symptoms of PTSD explain the association between low-level MOB exposure and PCS. As such, we cannot recommend specific strategies for mitigating the neurological effects of low-level MOB, because these effects have not yet been established. If, however, it is assumed that neurological effects of low-level MOB exposure are significant, successful intervention will require knowledge regarding effective prevention strategies. To target research in this area requires an understanding of the most-frequent intermediate- and long-term risks. It is clear that more research is needed to understand whether low-level MOB exposure is a risk to service members, the specific risks and outcomes involved, and how to determine the best candidates for emerging preventive measures. In the ensuing paragraphs, gaps in the literature are discussed and we make related recommendations.

Among animals, studies in mouse and rat models suggest it is plausible that low-level MOB exposure could result in neurological effects. It is uncertain how blast exposure levels tested in animal studies relate to exposure levels in humans. Furthermore, many of these studies only involve a single blast exposure; therefore, it is unknown if the results of these studies would vary should there be multiple exposures. If further animal testing is to be done to assess the neurological effects of low-level MOB exposure, it should be done in larger animals, ideally nonhuman primates. Additionally, the cumulative effect of low-level blast exposures also needs further exploration within any future animal testing. Furthermore, testing is needed on larger animal species to determine the level, or range of levels, of psi that constitute safe and acceptable MOB exposure levels, and which levels lead to adverse outcomes (e.g., persistent neurocognitive or neuropathological changes; mild, moderate, and severe TBI). This is not yet defined in the literature, a fact that complicated our efforts to appropriately scope this review.

Among humans, completing carefully designed prospective, longitudinal research is essential. These studies should include a representative sample of service members anticipated to have varying levels of exposure to low-level MOB (records should include the intensity of the blast and number of exposures), appropriate accounting of potential confounding variables (e.g., past TBI and PTSD), validated measures of low-level blast exposure and key outcomes of interest, and concerted efforts to maximize follow-up during and after military service. Further work is needed to validate measures of exposure. Wearable sensors have the potential to give a more complete, reliable, and valid measure of MOB exposure, and collecting the data from those monitors would be a minimal burden to the service members wearing them.

However, sensor technology will need further work to ensure that the sensors could offer the previously mentioned measurement features and would be durable enough for military use. To accomplish this, we first recommend a long-term data-collection effort to measure exposure levels from MOB events in training. This effort should last for a period of three to five years. This collection can be performed with existing blast-exposure sensors, such as the Blast Gauge system. Training data (e.g., the type of heavy weapons training that a service member undergoes, recording the number of heavy-weapon launches during the exposure period) and any mental health or neurological issues should also be recorded. Second, we recommend a modeling effort to improve or develop precise mathematical models that link sensor data to overpressure forces on the brain.

Once this research is complete, sensors can be used to measure exposure in a longitudinal study. At the conclusion of that study, further research can be performed to analyze the collected data, apply and improve mathematical modeling to convert sensor data into measurements of overpressure forces on human brain and other tissues, and link the effects of multiple, low-level MOB exposure events to neurological health. Sensors could be used as an early detection device to signal potentially harmful levels of MOB exposure.

Finally, there is evidence that improvements to helmets, enforcing adherence to hearing protection, and supplemental antioxidants may eventually serve to mitigate the neurological effects of blast exposure (Moss, King, and Blackman, 2009; Oishi and Schacht, 2011; Stephenson and Stephenson, 2011; Ganpule et al., 2012). However, these have only been tested at higher-intensity blast exposure levels, and more research is required to determine (1) whether protection from low-level MOB exposure is necessary and (2) if these approaches are effective for mitigating the neurological effects of low-level MOB exposure.

We are not suggesting the abandonment of current protective measures against high-intensity combat blast injuries (e.g., mild, moderate, and severe TBI). However, as it pertains to low-level MOB exposure, the state of the science is preliminary. Implementing aggressive preventive programs against this threat—without adequate evidence of preventable injury—risks unintended consequences and the consumption of considerable resources without commensurate benefit.

## Overall Recommendations

Perhaps the most striking finding from this review of the literature is how little research has been completed to determine the organizational threat and service member health impact of low-level MOB exposure, in contrast with our rapidly improving empirical research base relating to blast-related traumatic (concussive) brain injury. A Center for New American Security review of low-level MOB exposures (Fish and Scharre, 2018), published as we were completing this review, came to similar conclusions regarding the level of scientific uncertainty that exists relating to the neurological effects of low-level MOB exposure and recommended aggressive, ongoing surveillance of both exposure and neurological outcomes. However, military service–wide efforts to implement population exposure surveillance are more likely to be ineffective if they are implemented before the neurological effects of low-level MOB exposure are better characterized.

Prior to rolling out population-based military surveillance efforts, epidemiologic and other research is needed to establish whether low-level MOB exposure poses neurological or other health risks to service members and what, if any, the specific risks are. Surveillance programs require specific information to ensure collection of appropriate data, information that does not appear to be known in sufficient detail yet. Additional research, prior to surveillance, can ensure that surveillance efforts are calibrated to assess the most-important aspects of blast exposure and related outcomes.

Currently, most data from human and animal studies address the effects of higher-level blast exposures that can lead to concussion and more severe TBI. These studies do not allow a parsing of low-level MOB exposure effects, and therefore cannot adequately inform understanding of the risk that low-level MOB exposure creates for service members. Therefore, our main recommendation is for research that advances understanding of the specific health effects of low-level MOB exposure. This should include epidemiological studies to determine the magnitude of the problem and longitudinal research to better characterize short-, intermediate-, and long-term health risks. Other studies can help to calibrate future surveillance efforts. For example, neuropsychological research should be completed to determine which assessments are best for early detection of the potential cognitive effects of repeated low-level MOB exposure. Similarly, animal research can help researchers to better conduct human studies. These studies should include nonhuman primate research, so that results will more easily translate to humans. Other recommendations include the development and testing of preventive interventions and biomarker and biosensor validation studies suited for use with preventive materiel and strategies.

# Full Text Articles Screened for Inclusion

Abbotts, Rachek, Stuart Edward Harrison, and Glenn L. Cooper, "Primary Blast Injuries to the Eye: A Review of the Evidence," *Journal of the Royal Army Medical Corps*, Vol. 153, No. 2, 2007, pp. 119–123.

Abdollahi, Abbas, Mansor Abu Talib, Siti Nor Yaacob, and Zolhabri Ismail, "Hardiness as a Mediator Between Perceived Stress and Happiness in Nurses," *Journal of Psychiatric and Mental Health Nursing*, Vol. 21, No. 9, November 2014, pp. 789–796.

Abe, Nobuhito, and Joshua D. Greene, "Response to Anticipated Reward in the Nucleus Accumbens Predicts Behavior in an Independent Test of Honesty," *Journal of Neuroscience*, Vol. 34, No. 32, August 6, 2014, pp. 10564–10572.

Ahlers, Stephen Thomas, Elaina Vasserman-Stokes, Michael Christopher Shaughness, Aaron Andrew Hall, Deborah Ann Shear, Mikulas Chavko, Richard Michael McCarron, and James Radford Stone, "Assessment of the Effects of Acute and Repeated Exposure to Blast Overpressure in Rodents: Toward a Greater Understanding of Blast and the Potential Ramifications for Injury in Humans Exposed to Blast," *Frontiers in Neurology*, Vol. 3, Article 32, March 2012.

Ari, Adrienne B., "Eye Injuries on the Battlefields of Iraq and Afghanistan: Public Health Implications," *Optometry–Journal of the American Optometric Association*, Vol. 77, No. 7, 2006, pp. 329–339.

Baalman, Kelli L., R James Cotton, S. Neil Rasband, and Matthew N. Rasband, "Blast Wave Exposure Impairs Memory and Decreases Axon Initial Segment Length," *Journal of Neurotrauma*, Vol. 30, No. 9, May 2013, pp. 741–751.

Bahraini, Nazanin H., Lisa A. Brenner, Jeri E. F. Harwood, Beeta Y. Homaifar, Susan E. Ladley-O'Brien, Christopher M. Filley, James P. Kelly, and Lawrence E. Adler, "Utility of the Trauma Symptom Inventory for the Asssessment of Post-Traumatic Stress Symptoms in Veterans with a History of Psychological Trauma and/or Brain Injury," *Military Medicine*, Vol. 174, No. 10, October 2009, pp. 1005–1009.

Barzilai, Liranr, Moti Harats, Itay Wiser, Oren Weissman, Noam Domniz, Elon Glassberg, Demetris Stavrou, Issac Zilinsky, Eyal Winkler, and Josef Hiak, "Characteristics of Improvised Explosive Device Trauma Casualties in the Gaza Strip and Other Combat Regions: The Israeli Experience," *Wounds*, Vol. 27, No. 8, August 2015, pp. 209–214.

Bir, Cynthia, *Measuring Blast-Related Intracranial Pressure Within the Human Head*, Detroit, Mich.: Wayne State University, 2011.

"Blast Overpressure Is Generated from the Firing of Weapons, and May Cause Brain Injury," *ScienceDaily*, January 20, 2009. As of July 18, 2019:
https://www.sciencedaily.com/releases/2009/01/090119091112.htm

Blennow, Kaj, Michael Jonsson, Niels Peter Andreasen, Lars Erik Rosengren, Anders Wallin, Pekka A. Hellström, and Henrik Zetterberg, "No Neurochemical Evidence of Brain Injury After Blast Overpressure by Repeated Explosions or Firing Heavy Weapons," *Acta Neurologica Scandinavica*, Vol. 123, No. 4, April 2011, pp. 245–251.

Breeze, John, L. S. Allanson-Bailey, N. C. Hunt, Mark J. Midwinter, A. E. Hepper, A. Monaghan, and A. J. Gibbons, "Surface Wound Mapping of Battlefield Occulo-Facial Injury," *Injury*, Vol. 43, No. 11, 2012, pp. 1856–1860.

Brennan, K. C., Dan Kaufmann, Punam Sawant, and Jeremy Theriot, *Dissecting the Roles of Brain Injury and Combat-Related Stress in Post-Traumatic Headache*, Salt Lake City, Utah: University of Utah, 2015.

Brody, David L., *Radiological-Pathological Correlations Following Blast-Related Traumatic Brain Injury in the Whole Human Brain Using Ex Vivo Diffusion Tensor Imaging*, St. Louis, Mo.: Washington University in St. Louis, January 2014.

Brody, David L., *Advanced MRI in Acute Military TBI*, St. Louis, Mo.: Washington University in Saint Louis, November 2015.

Brody, David L., and Christine Mac Donald, *Advanced MRI in Blast-Related TBI*, St. Louis, Mo.: Washington University in St. Louis, September 2009.

Budde, Matthew D., Alok Shah, Michael McCrea, William E. Cullinan, Frank A. Pintar, and Brian D. Stemper, "Primary Blast Traumatic Brain Injury in the Rat: Relating Diffusion Tensor Imaging and Behavior," *Frontiers in Neurology*, Vol. 4, Article 154, October 2013.

Cao, Jiqing, Farida El Gaamouch, James S. Meabon, Kole D. Meeker, Li Zhu, Margaret B. Zhong, John Bendik, Gregory Elder, Ping Jing, Jiahong Xia, Wenjie Luo, David G. Cook, and Dongming Cai, "ApoE4-Associated Phospholipid Dysregulation Contributes to Development of Tau Hyper-Phosphorylation After Traumatic Brain Injury," *Scientific Reports*, Vol. 7, September 2017.

Capo-Aponte, José E., Gina M. Jurek, David V. Walsh, Leonard A. Temme, William A. Ahroon, and Daniel W. Riggs, "Effects of Repetitive Low-Level Blast Exposure on Visual System and Ocular Structures," *Journal of Rehabilitation Research and Development*, Vol. 52, No. 3, 2015, pp. 273-290.

Capo-Aponté, Jose E., *Pupillometry and Saccades as Objective mTBI Biomark*, Tacoma, Wash.: Geneva Foundation, 2015.

Carr, Walter, Elena Polejaeva, Anna Grome, Beth Crandall, Christina LaValle, Stephanie Eonta, and Lee Young, "Relation of Repeated Low-Level Blast Exposure with Symptomology Similar to Concussion," *Journal of Head Trauma Rehabilitation*, Vol. 30, No. 1, January–February 2015, pp. 47–55.

Carr, Walter, James R. Stone, Tim Walilko, Lee Ann Young, Tianlu Li Snook, Michelle E. Paggi, Jack W. Tsao, Christopher J. Jankosky, Robert V. Parish, and Stephen T. Ahlers, "Repeated Low-Level Blast Exposure: A Descriptive Human Subjects Study," *Military Medicine*, Vol. 181, Supplement 5, May 2016, pp. 28–39.

Casto, Kristen L., and Amy E. Nedostup, *Auditory, Vestibular and Cognitive Effects Due to Repeated Blast Exposure on the Warfighter*, Lakewood, Wash.: Geneva Foundation, July, 2012.

Chavko, Mikulas, Tomas Watanabe, Saleena Adeeb, Jason Lankasky, Stephen T. Ahlers, and Richard M. McCarron, "Relationship Between Orientation to a Blast and Pressure Wave Propagation Inside the Rat Brain," *Journal of Neuroscience Methods*, Vol. 195, No. 1, January 2011, pp. 61–66.

Choi, Jae Hyek, Whitney A. Greene, Anthony J. Johnson, Mikulas Chavko, Jeffery M. Cleland, Richard M. McCarron, and Heuy-Ching Wang, "Pathophysiology of Blast-Induced Ocular Trauma in Rats After Repeated Exposure to Low-Level Blast Overpressure," *Clinical and Experimental Ophthalmology*, Vol. 43, No. 3, April 2015, pp. 239–246.

Coats, Brittany, and Daniel F. Shedd, "Biomechanics of Eye Injury in the Military," in Amit Gefen and Yoram Epstein, eds., *The Mechanobiology and Mechanophysiology of Military-Related Injuries*, New York: Springer, 2015, pp. 235–262.

Coats, Brittany, and Daniel Shedd, *Temporal Progression of Visual Injury from Blast Exposure*, Salt Lake City, Utah: University of Utah, 2016. As of July 18, 2019:
http://www.dtic.mil/dtic/tr/fulltext/u2/1020693.pdf

Cockerham, Glenn C., Thomas A. Rice, Eva H. Hewes, Kimberly P. Cockerham, Sonne Lemke, Gloria Wang, Richard C. Lin, Catherine Glynn-Milley, and Lars Zumhagen, "Closed-Eye Ocular Injuries in the Iraq and Afghanistan Wars," *New England Journal of Medicine*, Vol. 364, No. 22, June 2011, pp. 2172–2173.

Committee on Gulf War and Health, *Gulf War and Health*, Vol. 7: *Long-Term Consequences of Traumatic Brain Injury*, Washington, D.C.: The National Academies Press, 2008.

Committee on Gulf War and Health, *Gulf War and Health*, Vol. 9: *Long-Term Effects of Blast Exposures*, Washington, D.C.: The National Academies Press, 2014.

Committee on Infectious Diseases, American Academy of Pediatrics, "Infection Prevention and Control in Pediatric Ambulatory Settings," *Pediatrics*, Vol. 120, No. 3, Sep, 2007, pp. 650–665.

Concannon, Thomas W., Melissa Fuster, Tully Saunders, Kamal Patel, John B. Wong, Laurel K. Leslie, and Joseph Lau, "A Systematic Review of Stakeholder Engagement in Comparative Effectiveness and Patient-Centered Outcomes Research," *Journal of General Internal Medicine*, Vol. 29, No. 12, December 2014, pp. 1692–1701.

Cornis-Pop, Micaela, Pauline A. Mashima, Carole R. Roth, Donald L. MacLennan, Linda M. Picon, Carol Smith Hammond, Shari Goo-Yoshino, Emi Isaki, Maile Singson, and Elaine M. Frank, "Cognitive-Communication Rehabilitation for Combat-Related Mild Traumatic Brain Injury," *Journal of Rehabilitation Research and Development*, Vol. 49, No. 7, 2012, pp. 11–31.

De Laet, C., Georges Casimir, Jean Duchateau, Esther Vamos, Christine Devalck, E. Sariban, and A. Ferster, "Leukemia Lymphoma T-Cell as First Manifestation of Ataxia-Telangiectasia," *Archives de Pediatrie: Organe Officiel de la Societe Francaise de Pediatrie*, Vol. 3, No. 7, 1996, pp. 681–684.

Delano-Wood, Lisa, *Quantitative Tractography and Volumetric MRI in Blast and Blunt Force TBI: Predictors of Neurocognitive and Behavioral Outcome*, San Diego, Calif.: Veterans Medical Research Foundation, October 2014.

DeMar, James C, *Elucidation of Inflammation Processes Exacerbating Neuronal Cell Damage to the Retina and Brain Visual Centers as Quest for Therapeutic Drug Targets in Rat Model of Blast Overpressure Wave Exposure*, Tacoma, Wash.: Geneva Foundation, October 2016.

DeMunck, Casey G., *Baseline Establishment Using Virtual Environment Traumatic Brain Injury Screen (VETS)*, thesis, Monterey, Calif.: Naval Postgraduate School, June 2015. As of July 18, 2019: http://www.dtic.mil/dtic/tr/fulltext/u2/a632324.pdf

Department of Defense Blast Injury Research Program Coordinating Office, *Biomedical Basis for Mild Traumatic Brain Injury (mTBI) Environmental Sensor Threshold Values*, McLean, Va., 2014. As of July 18, 2019: https://blastinjuryresearch.amedd.army.mil/assets/docs/sos/meeting_proceedings/2014_SoS_Meeting_Proceedings.pdf

Dougherty, Amber L., Andrew J. MacGregor, Peggy P. Han, Erik S. Viirre, Kevin J. Heltemes, and Michael R. Galarneau, "Blast-Related Ear Injuries Among U.S. Military Personnel," *Journal of Rehabilitation Research and Development*, Vol. 50, No. 6, November 2013, pp. 893–904.

Dretsch, Michael, *Evaluation of the King-Devick Test to Assess Eye Movements and the Performance of Rapid Number Naming in Concussed and Non-Concussed Service Members*, Tacoma, Wash.: Geneva Foundation, July 2017.

Elder, Gregory A., Nathan P. Dorr, Rita De Gasperi, Miguel A. Gama Sosa, Michael C. Shaughness, Eric Maudlin-Jeronimo, Aaron A. Hall, Richard M. McCarron, and Stephen T. Ahlers, "Blast Exposure Induces Post-Traumatic Stress Disorder-Related Traits in a Rat Model of Mild Traumatic Brain Injury," *Journal of Neurotrauma*, Vol. 29, No. 16, 2012, pp. 2564–2575.

Elder, Gregory A., James R. Stone, and Stephen T. Ahlers, "Effects of Low-Level Blast Exposure on the Nervous System: Is There Really a Controversy?" *Frontiers in Neurology*, Vol. 5, Article 269, December 2014.

Ettenhofer, Mark L., Rebecca J. Melrose, Zainab Delawalla, Steven A. Castellon, and Anna Okonek, "Correlates of Functional Status Among OEF/OIF Veterans with a History of Traumatic Brain Injury," *Military Medicine*, Vol. 177, No. 11, November 2012, p. 1272–1278.

Evans, Charlesnika T., Justin R. St. Andre, Theresa L.-B. Pape, Monica L. Steiner, Kevin T. Stroupe, Timothy P. Hogan, Frances M. Weaver, and Bridget M. Smith, "An Evaluation of the Veterans Affairs Traumatic Brain Injury Screening Process Among Operation Enduring Freedom and/or Operation Iraqi Freedom Veterans," *Physical Medicine and Rehabilitation*, Vol. 5, No. 3, March 2013, pp. 210–220.

Eve, David J., Martin R. Steele, Paul R. Sanberg, and Cesar V. Borlongan, "Hyperbaric Oxygen Therapy as a Potential Treatment for Post-Traumatic Stress Disorder Associated with Traumatic Brain Injury," *Neuropsychiatric Disease and Treatment*, Vol. 12, October 2016, pp. 2689–2705.

Ewert, Donald L., Jianzhong Lu, Wei Li, Xiaoping Du, Robert Floyd, and Richard Kopke, "Antioxidant Treatment Reduces Blast-Induced Cochlear Damage and Hearing Loss," *Hearing Research*, Vol. 285, No. 1–2, March 2012, pp. 29–39.

Farmer, Carrie M., Heather Krull, Thomas W. Concannon, Molly Simmons, Francesca Pillemer, Teague Ruder, Andrew Parker, Maulik P. Purohit, Liisa Hiatt, and Benjamin Saul Batorsky, "Understanding Treatment of Mild Traumatic Brain Injury in the Military Health System," *RAND Health Quarterly*, Vol. 6, No. 2, 2017. As of July 18, 2019: https://www.rand.org/pubs/periodicals/health-quarterly/issues/v6/n2/11.html

Fausti, Stephen A., Debra J. Wilmington, Frederick J. Gallun, Paula J. Myers, and James A. Henry, "Auditory and Vestibular Dysfunction Associated with Blast-Related Traumatic Brain Injury," *Journal of Rehabilitation Research and Development*, Vol. 46, No. 6, 2009, pp. 797–809.

Fear, Nicola T., Daniel Meek, Paul Cawkill, Norman Jones, Neil Greenberg, and Simon Wessely, "The Health of UK Civilians Deployed to Iraq," *European Journal of Public Health*, Vol. 27, No. 2, April 2017, pp. 367–371.

Fischer, Barbara L., Michael Parsons, Sally Durgerian, Christine Reece, Lyla Mourany, Mark J. Lowe, Erik B. Beall, Katherine A. Koenig, Stephen E. Jones, Mary R. Newsome, Randall S. Scheibel, Elisabeth A. Wilde, Maya Troyanskaya, Tricia L. Merkley, Mark Walker, Harvey S. Levin, and Stephen M. Rao, "Neural Activation During Response Inhibition Differentiates Blast from Mechanical Causes of Mild to Moderate Traumatic Brain Injury," *Journal of Neurotrauma*, Vol. 31, No. 2, 2014, pp. 169–179.

Fish, Lauren, and Paul Scharre, *Protecting Warfighters from Blast Injury*, Washington, D.C.: Center for a New American Security, 2018.

Fiskum, Gary, and William Fourney, *Underbody Blast Models of TBI Caused by Hyper-Acceleration and Secondary Head Impact*, Baltimore, Md.: University of Maryland, February 2014.

Forbes, David, Susan Fletcher, Andrea Phelps, Darryl Wade, Mark Creamer, and Meaghan O'Donnell, "Impact of Combat and Non-Military Trauma Exposure on Symptom Reduction Following Treatment for Veterans with Posttraumatic Stress Disorder," *Psychiatry Research*, Vol. 206, No. 1, March 2013, pp. 33–36.

Fortier, Catherine Brawn, Melissa M. Amick, Laura Grande, Susan McGlynn, Alexandra Kenna, Lindsay Morra, Alexandra Clark, William P. Milberg, and Regina E. McGlinchey, "The Boston Assessment of Traumatic Brain Injury-Lifetime (BAT-L) Semistructured Interview: Evidence of Research Utility and Validity," *Journal of Head Trauma Rehabilitation*, Vol. 29, No. 1, January–February 2014, pp. 89–98.

Friedl, Karl E., Stephen J. Grate, Susan P. Proctor, James W. Ness, Brian J. Lukey, and Robert L. Kane, "Army Research Needs for Automated Neuropsychological Tests: Monitoring Soldier Health and Performance Status," *Archives of Clinical Neuropsychology*, Vol. 22, Supplement 1, February 2007, pp. S7–S14.

Gaines, Katy D., Henry V. Soper, and Gholam R. Berenji, "Executive Functioning of Combat Mild Traumatic Brain Injury," *Applied Neuropsychology-Adult*, Vol. 23, No. 2, March 2016, pp. 115–124.

Gallun, Frederick J., M. Samantha Lewis, Robert L. Folmer, Anna C. Diedesch, Lina R. Kubli, Daniel J. McDermott, Therese C. Walden, Stephen A. Fausti, Henry L. Lew, and Marjorie R. Leek, "Implications of Blast Exposure for Central Auditory Function: A Review," *Journal of Rehabilitation Research and Development*, Vol. 49, No. 7, 2012, pp. 1059–1074.

Gama Sosa, Miguel A., Rita De Gasperi, Pierce L. Janssen, Frank J. Yuk, Pameka C. Anazodo, Paul E. Pricop, Alejandro J. Paulino, Bridget A. Wicinski, Michael Christopher Shaughness, Eric Maudlin-Jeronimo, Aaron A. Hall, Dara L. Dickstein, Richard M. McCarron, Mikulas Chavko, Patrick R. Hof, Stephen T. Ahlers, and Gregory A. Elder, "Selective Vulnerability of the Cerebral Vasculature to Blast Injury in a Rat Model of Mild Traumatic Brain Injury," *Acta Neuropathologica Communications*, Vol. 2, No. 67, 2014.

Gama Sosa, Miguel A., Rita De Gasperi, Alejandro J. Paulino, Paul E. Pricop, Michael C. Shaughness, Eric Maudlin-Jeronimo, Aaron A. Hall, William G. M. Janssen, Frank J. Yuk, Nathan P. Dorr, Dara L. Dickstein, Richard M. McCarron, Mikulas Chavko, Patrick R. Hof, Stephen T. Ahlers, and Gregory A. Elder, "Blast Overpressure Induces Shear-Related Injuries in the Brain of Rats Exposed to a Mild Traumatic Brain Injury," *Acta Neuropathologica Communications*, Vol. 1, No. 51, 2013.

Gan, Rong, *Biomechanical Modeling and Measurement of Blast Injury and Hearing Protection Mechanisms*, Norman, Okla.: University of Oklahoma, Norman, 2015.

Gandy, Sam, Milos D. Ikonomovic, Effie Mitsis, Gregory Elder, Stephen T. Ahlers, Jeffrey Barth, James R. Stone, and Steven T. DeKosky, "Chronic Traumatic Encephalopathy: Clinical-Biomarker Correlations and Current Concepts in Pathogenesis," *Molecular Neurodegeneration*, Vol. 9, No. 37, September 17, 2014.

Ganpule, Shailesh G., Linxia Gu, Aaron Alai, and Namas Chandra, "Role of Helmet in the Mechanics of Shock Wave Propagation Under Blast Loading Conditions," *Computer Methods in Biomechanics and Biomedical Engineering*, Vol. 15, No. 11, 2012, pp. 1233–1244.

Gardner, Raquel C., and Kristine Yaffe, "Epidemiology of Mild Traumatic Brain Injury and Neurodegenerative Disease," *Molecular and Cellular Neuroscience*, Vol. 66, May 2015, pp. 75–80.

Geisert, Eldon E., *Genetic Networks Activated by Blast Injury to the Eye*, Atlanta, Ga.: Emory University, August 2015.

Genovese, Raymond F., Jitendra Dave, and Stephen Ahlers, *Neurocognitive and Biomarker Evaluation of Combination mTBI from Blast Overpressure and Traumatic Stress*, Tacoma, Wash.: Geneva Foundation, 2014.

Goswami, Ruma, Paul A. Dufort, Maria Carmela Tartaglia, Robin E. A. Green, A. Crawley, C. H. Tator, R. Wennberg, David J. Mikulis, Michelle Keightley, and Karen D. Davis, "Frontotemporal Correlates of Impulsivity and Machine Learning in Retired Professional Athletes with a History of Multiple Concussions," *Brain Structure and Function*, Vol. 221, No. 4, May 2016, pp. 1911–1925.

Goverover, Yael, and Nancy Chiaravalloti, "The Impact of Self-Awareness and Depression on Subjective Reports of Memory, Quality-of-Life and Satisfaction with Life Following TBI," *Brain Injury*, Vol. 28, No. 2, 2014, pp. 174–180.

Gray, Walt, Matthew Reilly, Brian J. Lund, Jae H. Choi, William E. Sponsel, and Randolph D. Glickman, *Sub-Lethal Ocular Trauma (SLOT): Establishing a Standardized Blast Threshold to Facilitate Diagnostic, Early Treatment, and Recovery Studies for Blast Injuries to the Eye and Optic Nerve*, San Antonio, Tex.: University of Texas, 2015.

Green, Kimberly T., Jean C. Beckham, Nagy Youssef, and Eric B. Elbogen, "Alcohol Misuse and Psychological Resilience Among U.S. Iraq and Afghanistan-Era Veteran Military Personnel," *Addictive Behaviors*, Vol. 39, No. 2, February 2014, pp. 406–413.

Green, Steven, *Prevention of Noise Damage to Cochlear Synapses*, Iowa City, Ia.: University of Iowa, 2016.

Groenewold, Matthew R., Elizabeth A. Masterson, Christa L. Themann, and Rickie R. Davis, "Do Hearing Protectors Protect Hearing?" *American Journal of Industrial Medicine*, Vol. 57, No. 9, September 2014, pp. 1001–1010.

Grujicic, M. Danica, William Cameron Bell, Bhaskar Pandurangan, and Patrick Glomski, "Fluid/Structure Interaction Computational Investigation of Blast-Wave Mitigation Efficacy of the Advanced Combat Helmet," *Journal of Materials Engineering and Performance*, Vol. 20, No. 6, 2010, pp. 877–893.

Grujicic, M. Danica, W. C. Bell, B. Pandurangan, and T. He, "Blast-Wave Impact-Mitigation Capability of Polyurea When Used as Helmet Suspension-Pad Material," *Materials and Design*, Vol. 31, No. 9, 2010, pp. 4050–4065.

Gunther, Peter J., and Mark S. Riddle, "Effect of Combat Eye Protection on Field of View Among Active-Duty U.S. Military Personnel," *Optometry–Journal of the American Optometric Association*, Vol. 79, No. 11, November 2008, pp. 663–669.

Gupta, Raj K., and Andrzej Przekwas, "Mathematical Models of Blast-Induced TBI: Current Status, Challenges, and Prospects," *Frontiers in Neurology*, Vol. 4, Article 59, May 30, 2013.

Hamilton, Jon, "Do U.S. Troops Risk Brain Injury When They Fire Heavy Weapons?" *National Public Radio*, April 5, 2017. As of October 1, 2017:
https://www.npr.org/sections/health-shots/2017/04/05/522613294/
do-u-s-troops-risk-brain-injury-when-they-fire-heavy-weapons

Han, Kihwan, Christine L. Mac Donald, Ann M. Johnson, Yolanda Barnes, Linda Wierzechowski, David Zonies, John Oh, Stephen Flaherty, Raymond Fang, and Marcus E. Raichle, "Disrupted Modular Organization of Resting-State Cortical Functional Connectivity in U.S. Military Personnel Following Concussive 'Mild' Blast-Related Traumatic Brain Injury," *NeuroImage*, Vol. 84, January 2014, pp. 76–96.

Haran, F. J., Aimee L. Alphonso, Alia Creason, Justin S. Campbell, Dagny Johnson, Emily Young, and Jack W. Tsao, "Analysis of Post-Deployment Cognitive Performance and Symptom Recovery in U.S. Marines," *PLoS One*, Vol. 8, No. 11, 2013, e79595.

Haran, F. J., Aimee L. Alphonso, Alia Creason, Justin S. Campbell, Dagny Johnson, Emily Young, and Jack W. Tsao, "Reliable Change Estimates for Assessing Recovery From Concussion Using the ANAM4 TBI-MIL," *Journal of Head Trauma Rehabilitation*, Vol. 31, No. 5, September–October 2016, pp. 329–338.

Hartings, Jed A., M. Ross Bullock, David O. Okonkwo, Lilian S. Murray, Gordon D. Murray, Martin Fabricius, Andrew I. R. Maas, Johannes Woitzik, Oliver Sakowitz, Bruce Mathern, Bob Roozenbeek, Hester Lingsma, Jens P. Dreier, Ava M. Puccio, Lori A. Shutter, Clemens Pahl, and Anthony J. Strong, "Spreading Depolarisations and Outcome After Traumatic Brain Injury: A Prospective Observational Study," *Lancet Neurology*, Vol. 10, No. 12, December 2011, pp. 1058–1064.

Hayes, Jasmeet Pannu, Rajendra A. Morey, and Larry A. Tupler, "A Case of Frontal Neuropsychological and Neuroimaging Signs Following Multiple Primary-Blast Exposure," *Neurocase*, Vol. 18, No. 3, June 2012, pp. 258–269.

Heldt, Scott A., Andrea J. Elberger, Yunping Deng, Natalie Hart Guley, Nobel A. Del Mar, Joshua T. Rogers, Gy Won Choi, Jessica Ferrell, Tonia S. Rex, Marcia G. Honig, and Anton Reiner, "A Novel Closed-Head Model of Mild Traumatic Brain Injury Caused by Primary Overpressure Blast to the Cranium Produces Sustained Emotional Deficits in Mice," *Frontiers in Neurology*, Vol. 5, Article 2, 2014.

Heller, Michael J., *Rapid Isolation and Detection for RNA Biomarkers for TBI Diagnostics*, La Jolla, Calif.: University of California, San Diego, 2016.

Hellewell, Sarah, Bridgette D. Semple, and Maria Cristina Morganti-Kossmann, "Therapies Negating Neuroinflammation After Brain Trauma," *Brain Research*, Vol. 1640, Part A, June 2016, pp. 36–56.

Helling, Eric Robert, "Otologic Blast Injuries Due to the Kenya Embassy Bombing," *Military Medicine*, Vol. 169, No. 11, November 2004, pp. 872–876.

Hernández, Theresa D., Lisa A. Brenner, Kristen H. Walter, Jill E. Bormann, and Birgitta Johansson, "Complementary and Alternative Medicine (CAM) Following Traumatic Brain Injury (TBI): Opportunities and Challenges," *Brain Research*, Vol. 1640, Part A, June 2016, pp. 139–151.

Ho, Kevin H., and James H. Stuhmiller, *A Health Hazard Assessment for Blast Overpressure Exposures Subtitle - Analysis of RFR Biological Effects*, San Diego, Calif.: Jaycor, June 1997.

Hobson, Jonathan, Edward Chisholm, and Amr El Refaie, "Sound Therapy (Masking) in the Management of Tinnitus in Adults," *Cochrane Database of Systematic Reviews*, No. 12, December 2010.

Hoge, Charles W., Dennis McGurk, Jeffrey L. Thomas, Anthony L. Cox, Charles C. Engel, and Carl A. Castro, "Mild Traumatic Brain Injury in U.S. Soldiers Returning from Iraq," *New England Journal of Medicine*, Vol. 358, No. 5, January 2008, pp. 453–463.

Huber, Bertrand R., Michael L. Alosco, Thor D. Stein, and Ann C. McKee, "Potential Long-Term Consequences of Concussive and Subconcussive Injury," *Physical Medicine and Rehabilitation Clinics of North America*, Vol. 27, No. 2, May 2016, pp. 503–511.

Huber, Bertrand R., James S. Meabon, Zachary S. Hoffer, Jing Zhang, Jake G. Hoekstra, Kathleen F. Pagulayan, Pamela J. McMillan, Cynthia L. Mayer, William A. Banks, Brian C. Kraemer, Murray A. Raskind, Dorian B. McGavern, Elaine R. Peskind, and David G. Cook, "Blast Exposure Causes Dynamic Microglial/Macrophage Responses and Microdomains of Brain Microvessel Dysfunction," *Neuroscience*, Vol. 319, 2016, pp. 206–220.

Huber, Bertrand R., James S. Meabon, Tobin J. Martin, Pierre D. Mourad, Raymond Bennett, Brian C. Kraemer, Ibolja Cernak, Eric C. Petrie, Michael J. Emery, Erik R. Swenson, Cynthia Mayer, Edin Mehic, Elaine R. Peskind, and David G. Cook, "Blast Exposure Causes Early and Persistent Aberrant Phospho- and Cleaved-Tau Expression in a Murine Model of Mild Blast-Induced Traumatic Brain Injury," *Journal of Alzheimer's Disease*, Vol. 37, No. 2, 2013, pp. 309–323.

Hue, Christopher D., Frances S. Cho, Siqi Q. Cao, Cameron R. Bass, David F. Meaney, and Barclay Morrison III, "Dexamethasone Potentiates in Vitro Blood-Brain Barrier Recovery after Primary Blast Injury by Glucocorticoid Receptor-Mediated Upregulation of ZO-1 Tight Junction Protein," *Journal of Cerebral Blood Flow and Metabolism*, Vol. 35, No. 7, July 2015, pp. 1191–1198.

Hue, Christopher D., Siqi Cao, Syed F. Haider, Kiet V. Vo, Gwen B. Effgen, Edward Vogel III, Matthew B. Panzer, Cameron R. Bass, David F. Meaney, and Barclay Morrison III, "Blood-Brain Barrier Dysfunction after Primary Blast Injury in Vitro," *Journal of Neurotrauma*, Vol. 30, No. 19, October 2013, pp. 1652–1663.

Ivanov, Iliyan, Corey Fernandez, Effie M. Mitsis, Dara L. Dickstein, Edmund Wong, Cheuk Y. Tang, Jessie Simantov, Charlene Bang, Erin Moshier, Mary Sano, Gregory A. Elder, and Erin A. Hazlett, "Blast Exposure, White Matter Integrity, and Cognitive Function in Iraq and Afghanistan Combat Veterans," *Frontiers in Neurology*, Vol. 8, 2017.

Iverson, Grant L., Jean A. Langlois, Michael A. McCrea, and James P. Kelly, "Challenges Associated with Post-Deployment Screening for Mild Traumatic Brain Injury in Military Personnel," *Clinical Neuropsychologist*, Vol. 23, No. 8, 2009, pp. 1299–1314.

Ivins, Brian J., Robert Kane, and Karen A. Schwab, "Performance on the Automated Neuropsychological Assessment Metrics in a Nonclinical Sample of Soldiers Screened for Mild TBI After Returning from Iraq and Afghanistan: A Descriptive Analysis," *Journal of Head Trauma Rehabilitation*, Vol. 24, No. 1, 2009, pp. 24–31.

Jaffee, Michael S., and Kimberly S. Meyer, "A Brief Overview of Traumatic Brain Injury (TBI) and Post-Traumatic Stress Disorder (PTSD) Within the Department of Defense," *Clinical Neuropsychologist*, Vol. 23, No. 8, 2009, pp. 1291–1298.

Jaffee, Michael S., Kathy M. Helmick, Philip D. Girard, Kim S. Meyer, Kathy Dinegar, and Karyn Hede George, "Acute Clinical Care and Care Coordination for Traumatic Brain Injury Within Department of Defense," *Journal of Rehabilitation Research and Development*, Vol. 46, No. 6, 2009, pp. 655–665.

Jankosky, Christopher J., Tomoko I. Hooper, Nisara S. Granado, Ann Scher, Gary D. Gackstetter, Edward J. Boyko, Tyler C. Smith, "Headache Disorders in the Millennium Cohort: Epidemiology and Relations with Combat Deployment," *Headache*, Vol. 51, No. 7, July/August 2011, pp. 1098–1111.

Job, Agnes, Pascal Hamery, Sebastien De Mezzo, J-C Fialaire, Andre A. le Roux, Micahel Untereiner, Federica Cardinale, Hugues Michel, Celine Klein, and Bill Belcourt, "Rifle Impulse Noise Affects Middle-Ear Compliance in Soldiers Wearing Protective Earplugs," *International Journal of Audiology*, Vol. 55, No. 1, 2016, pp. 30–37.

Johnson, Robin R., Djordje Popvic, Deborah Perlick, Dennis Dyck, and Chris Berka, "Development of Sensitive, Specific, and Deployable Methods for Detecting and Discriminating mTBI and PTSD," *International Conference on Foundations of Augmented Cognition*, 2009, pp. 826–835.

Jones, Alvin, "Test of Memory Malingering: Cutoff Scores for Psychometrically Defined Malingering Groups in a Military Sample," *Clinical Neuropsychologist*, Vol. 27, No. 6, August 2013, pp. 1043–1059.

Jones, Alvin, M. Victoria Ingram, and Yossef S. Ben-Porath, "Scores on the MMPI-2-RF Scales as a Function of Increasing Levels of Failure on Cognitive Symptom Validity Tests in a Military Sample," *Clinical Neuropsychologist*, Vol. 26, No. 5, 2012, pp. 790–815.

Jones, Kirstin, Jae-Hyek Choi, William E. Sponsel, Walt Gray, Sylvia L. Groth, Randolph D. Glickman, Brian J. Lund, and Matthew A. Reilly, "Low-Level Primary Blast Causes Acute Ocular Trauma in Rabbits," *Journal of Neurotrauma*, Vol. 33, No. 13, July 2016, pp. 1194–1201.

Joseph, Antony R., Jaime L. Horton, Mary C. Clouser, Andrew J. MacGregor, Michelle Louie, and Michael R. Galarneau, "Development of a Comprehensive Blast-Related Auditory Injury Database (BRAID)," *Journal of Rehabilitation Research and Development*, Vol. 53, No. 3, 2016, pp. 295–306.

Judge, John A., and Scott A. Mathews, *Fabrication and Testing of a Blast Concussion Burst Sensor*, Washington, D.C.: Catholic University of America, June 2009.

Kamimori, Gary H., L. A. Reilly, Christina R. LaValle, and U. B. Olaghere Da Silva, "Occupational Overpressure Exposure of Breachers and Military Personnel," *Shock Waves*, Vol. 27, No. 6, November 2017, pp. 837–847.

Kamins, Joshua, Erin Bigler, Tracey Covassin, Luke Henry, Simon Kemp, John J. Leddy, Andrew Mayer, Michael McCrea, Mayumi Prins, Kathryn J. Schneider, Tamara C. Valovich McLeod, Roger Zemek, and Christopher C. Giza, "What Is the Physiological Time to Recovery After Concussion? A Systematic Review," *British Journal of Sports Medicine*, Vol. 51, No. 12, 2017, pp. 935–940.

Kamnaksh, Alaa, Matthew D. Budde, Erzsebet Kovesdi, Joseph B. Long, Joseph A. Frank, and Denes V. Agoston, "Diffusion Tensor Imaging Reveals Acute Subcortical Changes After Mild Blast-Induced Traumatic Brain Injury," *Scientific Reports*, Vol. 4, 2014.

Kane, Michael J., Mariana Angoa-Perez, Denise I. Briggs, David C. Viano, Christian W. Kreipke, and Donald M. Kuhn, "A Mouse Model of Human Repetitive Mild Traumatic Brain Injury," *Journal of Neuroscience Methods*, Vol. 203, No. 1, January 15, 2012, pp. 41–49.

Karr, Justin E., Corson N. Areshenkoff, Emily C. Duggan, and Mauricio A. Garcia-Barrera, "Blast-Related Mild Traumatic Brain Injury: A Bayesian Random-Effects Meta-Analysis on the Cognitive Outcomes of Concussion Among Military Personnel," *Neuropsychology Review*, Vol. 24, No. 4, December 2014, pp. 428–444.

Kawoos, Usmah, Richard M. McCarron, and Mikulas Chavko, "Protective Effect of N-Acetylcysteine Amide on Blast-Induced Increase in Intracranial Pressure in Rats," *Frontiers in Neurology*, Vol. 8, Article 291, 2017.

Kelley, Amanda M., Bethany M. Ranes, Art Estrada, and Catherine M. Grandizio, "Evaluation of the Military Functional Assessment Program: Preliminary Assessment of the Construct Validity Using an Archived Database of Clinical Data," *Journal of Head Trauma Rehabilitation*, Vol. 30, No. 4, July–August 2015, pp. E11–20.

Kennedy, Carrie H., J. Porter Evans, Shawnna Chee, Jeffrey L. Moore, Jeffrey T. Barth, and Keith A. Stuessi, "Return to Combat Duty After Concussive Blast Injury," *Archives of Clinical Neuropsychology*, Vol. 27, No. 8, December 2012, pp. 817–827.

Klemenhagen, Kristen C., Scott P. O'Brien, and David L. Brody, "Repetitive Concussive Traumatic Brain Injury Interacts with Post-Injury Foot Shock Stress to Worsen Social and Depression-Like Behavior in Mice," *PLoS One*, Vol. 8, No. 9, September 2013, e74510.

Kluchinsky, Timothy A., Jr., Charles R. Jokel, John V. Cambre, Donald E. Goddard, and Robert W. Batts Jr., "The Health Hazard Assessment Process in Support of Joint Weapon System Acquisitions," *Army Medical Department Journal*, Vol. 53, No. 7, April–June 2013.

Kobeissy, Firas, Stefania Mondello, Nihal Tümer, Hale Z. Toklu, Melissa A. Whidden, Nataliya Kirichenko, Zhiqun Zhang, Victor Prima, Walid Yassin, John Anagli, Namas Chandra, Stan Sveltov, and Kevin K. W. Wang, "Assessing Neuro-Systemic and Behavioral Components in the Pathophysiology of Blast-Related Brain Injury," *Frontiers in Neurology*, Vol. 4, November 21, 2013.

Kochanek, Patrick M., C. Edward Dixon, David K. Shellington, Samuel S. Shin, Hülya Bayır, Edwin K. Jackson, Valerian E. Kagan, Hong Q. Yan, Peter V. Swauger, Steven A. Parks, David V. Ritzel, Richard Bauman, Robert S. B. Clark, Robert H. Garman, Faris Bandak, Geoffrey Ling, and Larry W. Jenkins, "Screening of Biochemical and Molecular Mechanisms of Secondary Injury and Repair in the Brain After Experimental Blast-Induced Traumatic Brain Injury in Rats," *Journal of Neurotrauma*, Vol. 30, No. 11, 2013, pp. 920–937.

Koliatsos, Vassilis E., Ibolja Cernak, Leyan Xu, Yeajin J. Song, Alena Savonenko, Barbara J. Crain, Charles G. Eberhart, Constantine E. Frangakis, Tatiana Melnikova, Hyunsu Kim, and Deidre Lee, "A Mouse Model of Blast Injury to Brain: Initial Pathological, Neuropathological, and Behavioral Characterization," *Journal of Neuropathology and Experimental Neurology*, Vol. 70, No. 5, May 2011, pp. 399–416.

Kontos, Anthony P., Russ S. Kotwal, R. J. Elbin, Robert H. Lutz, Robert D. Forsten, Peter J. Benson, and Kevin M. Guskiewicz, "Residual Eeffects of Combat-Related Mild Traumatic Brain Injury," *Journal of Neurotrauma*, Vol. 30, No. 8, 2013, pp. 680–686.

Kubli, Lina R., Robin L. Pinto, Holly L. Burrows, Philip D. Littlefield, and Douglas S. Brungart, "The Effects of Repeated Low-Level Blast Exposure on Hearing in Marines," *Noise and Health*, Vol. 19, No. 90, 2017, p. 227–238.

Kulbe, Jacqueline R., and James W. Geddes, "Current Status of Fluid Biomarkers in Mild Traumatic Brain Injury," *Experimental Neurology*, Vol. 275, Pt. 3, January 2016, pp. 334–352.

Kulkarni, Shilpa G., Xiang Gao, Stephen V. Horner, Ji Qing Zheng, and N. V. David, "Ballistic Helmets—Their Design, Materials, and Performance Against Traumatic Brain Injury," *Composite Structures*, Vol. 101, July 2013, pp. 313–331.

LaFiandra, Michael E., and Harry Zywiol, *Non-Line-of-Sight Cannon (NLOS-C) System Crew Shock Loading, Evaluation of Potential Head and Neck Injury*, Army Research Laboratory, ARL-TR-4228, August 2007.

Laksari, Kaveh, Lyndia C. Wu, Mehmet Kurt, Calvin Kuo, and David C. Camarillo, "Resonance of Human Brain Under Head Acceleration," *Journal of the Royal Society Interface*, Vol. 12, No. 108, July 6, 2015.

Lemke, Sonne, Glenn C. Cockerham, Catherine Glynn-Milley, and Kimberly P. Cockerham, "Visual Quality of Life in Veterans With Blast-Induced Traumatic Brain Injury," *JAMA Ophthalmology*, Vol. 131, No. 12, December 2013, pp. 1602–1609.

Levin, Harvey S., and Claudia S. Robertson, "Mild Traumatic Brain Injury in Translation," *Journal of Neurotrauma*, Vol. 30, No. 8, April 2013, pp. 610–617.

Lew, Henry L., James F. Jerger, Sylvia B. Guillory, and James A. Henry, "Auditory Dysfunction in Traumatic Brain Injury," *Journal of Rehabilitation Research and Development*, Vol. 44, No. 7, 2007, p. 921–928.

Lim, Yi Wei, Nathan P. Meyer, Alok S. Shah, Matthew D. Budde, Brian D. Stemper, and Christopher M. Olsen, "Voluntary Alcohol Intake Following Blast Exposure in a Rat Model of Mild Traumatic Brain Injury," *PLoS One*, Vol. 10, No. 4, April 2015.

Littlefield, Philip D., Robin L. Pinto, Holly L. Burrows, and Douglas S. Brungart, "The Vestibular Effects of Repeated Low-Level Blasts," *Journal of Neurotrauma*, Vol. 33, No. 1, 2016, pp. 71–81.

Liu, Xixia, Jianhua Qiu, Sasha Alcon, Jumana Hashim, William P. Meehan, and Rebekah Mannix, "Environmental Enrichment Mitigates Deficits after Repetitive Mild TBI," *Journal of Neurotrauma*, Vol. 34, No. 16, June 8, 2017.

Long, Joseph, *Combined Effects of Primary and Tertiary Blast on Rat Brain: Characterization of a Model of Blast-Induced Mild Traumatic Brain Injury*, Tacoma, Wash.: Geneva Foundation, 2013.

Long, Joseph, *Blast-Induced Acceleration in a Shock Tube: Distinguishing Primary and Tertiary Blast Injury*, Tacoma, Wash.: Geneva Foundation, 2016. As of July 18, 2019: http://www.dtic.mil/dtic/tr/fulltext/u2/1039065.pdf

Long, Joseph B., *Brain Vulnerability to Repeated Blast Overpressure and Polytrauma*, Tacoma, Wash.: Geneva Foundation, November 2013.

Long, Joseph B., *Kevlar Vest Protection Against Blast Overpressure Brain Injury: Systemic Contributions to Injury Etiology*, Tacoma, Wash.: Geneva Foundation, 2014.

Long, Joseph B., *Assessment and Treatment of Blast-Induced Auditory and Vestibular Injuries*, Tacoma, Wash.: Geneva Foundation, 2016.

Lucke-Wold, B. P., A. F. Logsdon, K. E. Smith, R. C. Turner, D. L. Alkon, Z. Tan, Z. J. Naser, C. M. Knotts, J. D. Huber, and C. L. Rosen, "Bryostatin-1 Restores Blood Brain Barrier Integrity Following Blast-Induced Traumatic Brain Injury," *Molecular Neurobiology*, Vol. 52, No. 3, December 2015, pp. 1119–1134.

Luo, Jian, Andy Nguyen, Saul Villeda, Hui Zhang, Zhaoqing Ding, Derek Lindsey, Gregor Bieri, Joseph M. Castellano, Gary S. Beaupre, and Tony Wyss-Coray, "Long-Term Cognitive Impairments and Pathological Alterations in a Mouse Model of Repetitive Mild Traumatic Brain Injury," *Frontiers in Neurology*, Vol. 5, Article 12, February 2014.

Mac Donald, Christine L., Ann M. Johnson, Dana Cooper, Elliot C. Nelson, Nicole J. Werner, Joshua S. Shimony, Abraham Z. Snyder, Marcus E. Raichle, John R. Witherow, Raymond Fang, Stephen F. Flaherty, and David L. Brody, "Detection of Blast-Related Traumatic Brain Injury in U.S. Military Personnel," *New England Journal of Medicine*, Vol. 364, No. 22, June 2011, pp. 2091–2100.

Macera, Caroline A., Hilary Jeanne Aralis, Andrew J. MacGregor, Mitchell J. Rauh, and Michael R. Galarneau, "Postdeployment Symptom Changes and Traumatic Brain Injury and/or Posttraumatic Stress Disorder in Men," *Journal of Rehabilitation Research and Development*, Vol. 49, No. 8, 2012, pp. 1197–1208.

MacGregor, Andrew J., Amber L. Dougherty, Rosemary Grace Hurford Morrison, Kimberly H. Quinn, and Michael R. Galarneau, "Repeated Concussion Among U.S. Military Personnel During Operation Iraqi Freedom," *Journal of Rehabilitation Research and Development*, Vol. 48, No. 10, 2011, pp. 1269–1278.

MacGregor, Andrew J., *Effects of Repeated Traumatic Brain Injuries in a Combat Setting*, San Diego, Calif.: Naval Health Research Center, December 2011.

Mader, Thomas H., Robert D. Carroll, Clifton S. Slade, Roger K. George, J. Phillip Ritchey, and S. Page Neville, "Ocular War Injuries of the Iraqi Insurgency, January–September 2004," *Ophthalmology*, Vol. 113, No. 1, January 2006, pp. 97–104.

Maguen, Shira, Erin Madden, Karen M. Lau, and Karen Seal, "The Impact of Head Injury Mechanism on Mental Health Symptoms in Veterans: Do Number and Type of Exposures Matter?" *Journal of Traumatic Stress*, Vol. 25, No. 1, February 2012, pp. 3–9.

Management of Concussion-mild Traumatic Brain Injury Working Group, *VA/DoD Clinical Practice Guideline for Management of Concussion-Mild Traumatic Brain Injury*, Washington, D.C.: U.S. Department of Veterans Affairs and U.S. Department of Defense, 2016. As of July 18, 2019: https://www.healthquality.va.gov/guidelines/Rehab/mtbi/

Manley, Geoffrey T., Amy J. Markowitz, and Brian Fabian, *TBI Endpoints Development*, San Francisco, Calif.: University of California, San Francisco, October 2015.

Maroon, Joseph C., Robert Winkelman, Jeffrey Bost, Austin Amos, Christina Mathyssek, and Vincent Miele, "Chronic Traumatic Encephalopathy in Contact Sports: A Systematic Review of All Reported Pathological Cases," *PLoS One*, Vol. 10, No. 2, February 2015.

McCulloch, Karen L., S. Goldman, Lynn Lowe, Mary V. Radomski, John Reynolds, R. Shapiro, and Therese A. West, "Development of Clinical Recommendations for Progressive Return to Activity After Military Mild Traumatic Brain Injury: Guidance for Rehabilitation Providers," *Journal of Head Trauma Rehabilitation*, Vol. 30, No. 1, January–February 2015, pp. 56–67.

McEntire, B. Joseph, V. Carol Chancey, Timothy Walilko, Gregory T. Rule, Gregory Weiss, Cameron Bass, and Jay Shridharani, *Helmet Sensor-Transfer Function and Model Development*, San Antonio, Tex.: True Research Foundation, 2010.

Meabon, James S., Bertrand R. Huber, Donna J. Cross, Todd L. Richards, Satoshi Minoshima, Kathleen F. Pagulayan, Ge Li, Kole D. Meeker, Brian C. Kraemer, Eric C. Petrie, Murray A. Raskind, Elaine R. Peskind, and David G. Cook, "Repetitive Blast Exposure in Mice and Combat Veterans Causes Persistent Cerebellar Dysfunction," *Science Translational Medicine*, Vol. 8, No. 321, 2016.

Merritt, Victoria C., Rael T. Lange, and Louis M. French, "Resilience and Symptom Reporting Following Mild Traumatic Brain Injury in Military Service Members," *Brain Injury*, Vol. 29, No. 11, 2015, pp. 1325–1336.

Miller, Greg, "Blast Injuries Linked to Neurodegeneration in Veterans," *Science*, Vol. 336, No. 6083, May 2012, pp. 790–791.

Moochhala, Shabbir M., Shirhan Md, Jia Lu, Choo-Hua Teng, and Colin Greengrass, "Neuroprotective Role of Aminoguanidine in Behavioral Changes After Blast Injury," *Journal of Trauma and Acute Care Surgery*, Vol. 56, No. 2, March 2004, pp. 393–403.

Morales-Tirado, Vanessa M., *Compound 49b Reduces Inflammatory Markers and Apoptosis After Ocular Blast Injury*, Memphis, Tenn.: University of Tennessee Health Science Center, November 2015.

Moss, William C., Michael J. King, and Eric G. Blackman, "Skull Flexure from Blast Waves: A Mechanism for Brain Injury with Implications for Helmet Design," *Physical Review Letters*, Vol. 103, No. 10, September 2009.

Mourad, Pierre D., *Toward Development of a Field-Deployable Imaging Device for TBI*, Seattle, Wash.: University of Washington, March 2013.

Newsome, Mary R., Sally Durgerian, Lyla Mourany, Randall S. Scheibel, Mark J. Lowe, Erik B. Beall, Katherine A. Koenig, Michael Parsons, Maya Troyanskaya, Christine Reece, Elisabeth Wilde, Barbara L. Fischer, Stephen E. Jones, Rajan Agarwal, Harvey S. Levin, and Stephen M. Rao, "Disruption of Caudate Working Memory Activation in Chronic Blast-Related Traumatic Brain Injury," *Neuroimage-Clinical*, Vol. 8, 2015, pp. 543–553.

Newsome, Mary R., Andrew R. Mayer, Xiaodi Lin, Maya Troyanskaya, George R. Jackson, Randall S. Scheibel, Annette Walder, Ajithraj Sathiyaraj, Elisabeth A. Wilde, Shalini Mukhi, Brian A. Taylor, and Harvey S. Levin, "Chronic Effects of Blast-Related TBI on Subcortical Functional Connectivity in Veterans," *Journal of the International Neuropsychological Society*, Vol. 22, No. 6, July 2016, pp. 631–642.

Norris, Jacob N., Scottie Smith, Erica Harris, David Walter Labrie, and Stephen T. Ahlers, "Characterization of Acute Stress Reaction Following an IED Blast-Related Mild Traumatic Brain Injury," *Brain Injury*, Vol. 29, No. 7–8, July 2015, pp. 898–904.

Nyein, Michelle K., Amanda M. Jason, Li Yu, Claudio M. Pita, John D. Joannopoulos, David F. Moore, and Raul A. Radovitzky, "In Silico Investigation of Intracranial Blast Mitigation with Relevance to Military Traumatic Brain Injury," *Proceedings of the National Academy of Sciences*, Vol. 107, No. 48, October 2010, pp. 20703–20708.

Oishi, Naoki, and Jochen Schacht, "Emerging Treatments for Noise-Induced Hearing Loss," *Expert Opinion on Emerging Drugs*, Vol. 16, No. 2, June 2011, pp. 235–245.

Okpala, Nnaemeka, "Knowledge and Attitude of Infantry Soldiers to Hearing Conservation," *Military Medicine*, Vol. 172, No. 5, May 2007, pp. 520–522.

Pagulayan, Kathleen F., Holly Rau, Renee Madathil, Madeleine Werhane, Steven P. Millard, Eric C. Petrie, Brett Parmenter, Sarah Peterson, Scott Sorg, Rebecca Hendrickson, Cindy Mayer, James S. Meabon, Bertrand R. Huber, Murray Raskind, David G. Cook, and Elaine R. Peskind, "Retrospective and Prospective Memory Among OEF/OIF/OND Veterans with a Self-Reported History of Blast-Related mTBI," *Journal of the International Neuropsychological Society*, Vol. 24, No. 4, April 2018, pp. 324–334.

Papa, Linda, "Potential Blood-Based Biomarkers for Concussion," *Sports Medicine and Arthroscopy Review*, Vol. 24, No. 3, September 2016, pp. 108–115.

Parish, R., W. Carr, Michelle Paggi, V. Anderson-Barnes, and M. Kelly, "The Neurocognitive Effect of Exposure to Repeated Low-Level Blasts in a Military Sample," *Archives of Clinical Neuropsychology*, Vol. 24, No. 5, August 2009, pp. 503–504.

Park, Eugene, Rebecca Eisen, Anna Kinio, and Andrew J. Baker, "Electrophysiological White Matter Dysfunction and Association with Neurobehavioral Deficits Following Low-Level Primary Blast Trauma," *Neurobiology of Disease*, Vol. 52, April 2013, pp. 150–159.

Park, Eugene, James J. Gottlieb, Bob Cheung, Pang Nin Shek, and Andrew James Baker, "A Model of Low-Level Primary Blast Brain Trauma Results in Cytoskeletal Proteolysis and Chronic Functional Impairment in the Absence of Lung Barotrauma," *Journal of Neurotrauma*, Vol. 28, No. 3, March 2011, pp. 343–357.

Peskind, Elaine R., Eric C. Petrie, Donna J. Cross, Kathleen Pagulayan, Kathleen McCraw, David Hoff, Kim Hart, Chang-En Yu, Murray A. Raskind, David G. Cook, and Satoshi Minoshima, "Cerebrocerebellar Hypometabolism Associated with Repetitive Blast Exposure Mild Traumatic Brain Injury in 12 Iraq War Veterans with Persistent Post-Concussive Symptoms," *Neuroimage*, Vol. 54, Supplement 1, January 2011, pp. S76–S82.

Petrie, Eric C., Donna J. Cross, Vasily L. Yarnykh, Todd Richards, Nathalie M. Martin, Kathleen Pagulayan, David Hoff, Kim Hart, Cynthia Mayer, Matthew Tarabochia, Murray A. Raskind, Satoshi Minoshima, and Elaine R. Peskind, "Neuroimaging, Behavioral, and Psychological Sequelae of Repetitive Combined Blast/Impact Mild Traumatic Brain Injury in Iraq and Afghanistan War Veterans," *Journal of Neurotrauma*, Vol. 31, No. 5, March 2014, pp. 425–436.

Por, Elaine D., Jae-Hyek Choi, and Brian J. Lund, "Low-Level Blast Exposure Increases Transient Receptor Potential Vanilloid 1 (TRPV1) Expression in the Rat Cornea," *Current Eye Research*, Vol. 41, No. 10, 2016, pp. 1294–1301.

Pun, Pamela Boon Li, Mary Kan, Agus Salim, Zhaohui Li, Kian Chye Ng, Shabbir M. Moochhala, Eng-Ang Ling, Mui Hong Tan, and Jia Lu, "Low Level Primary Blast Injury in Rodent Brain," *Frontiers in Neurology*, Vol. 2, Article 19, April 2011.

Quatman-Yates, Catherine, Amanda Cupp, Cherryanne Gunsch, Tonya Haley, Steve Vaculik, and David Kujawa, "Physical Rehabilitation Interventions for Post-mTBI Symptoms Lasting Greater than 2 Weeks: Systematic Review," *Physical Therapy*, Vol. 96, No. 11, November 3, 2016, pp. 1753–1763.

Rau, Holly K., Rebecca C. Hendrickson, Hannah C. Roggenkamp, Sarah Peterson, Brett Parmenter, David G. Cook, Elaine Peskind, and Kathleen F. Pagulayan, "Fatigue–But Not mTBI history, PTSD, or Sleep Quality–Directly Contributes to Reduced Prospective Memory Performance in Iraq and Afghanistan Era Veterans," *Clinical Neuropsychologist*, Vol. 32, No. 7, 2017, pp. 1–18.

Reid, Matthew W., Kelly J. Miller, Rael T. Lange, Douglas B. Cooper, David F. Tate, Jason Bailie, Tracey A. Brickell, Louis M. French, Sarah Asmussen, and Jan E. Kennedy, "A Multisite Study of the Relationships Between Blast Exposures and Symptom Reporting in a Post-Deployment Active Duty Military Population with Mild Traumatic Brain Injury," *Journal of Neurotrauma*, Vol. 31, No. 23, December 2014, pp. 1899–1906.

Rhea, Christopher K., Nikita A. Kuznetsov, Scott E. Ross, Benjamin Long, Jason T. Jakiela, Jason M. Bailie, Matthew A. Yanagi, F. Jay Haran, W. Geoffrey Wright, Rebecca K. Robins, Paul D. Sargent, and Joshua L. Duckworth, "Development of a Portable Tool for Screening Neuromotor Sequelae from Repetitive Low-Level Blast Exposure," *Military Medicine*, Vol. 182, Supplement 1, March 2017, pp. 147–154.

Riviere, Stephanie, Valerie Schwoebel, Karine Lapierre-Duval, Gerard Warret, Martine Saturnin, Paul Avan, Agnes Job, Thierry Lang, and The Expert Group, "Hearing Status After an Industrial Explosion: Experience of the AZF Explosion, 21 September 2001, France," *International Archives of Occupational and Environmental Health*, Vol. 81, No. 4, February 2008, pp. 409–414.

Roberts, Jennifer Carter, Timothy P. Harrigan, Emily E. Ward, Tammi M. Taylor, Melissa S. Annett, and Andrew C. Merkle, "Human Head–Neck Computational Model for Assessing Blast Injury," *Journal of Biomechanics*, Vol. 45, No. 16, November 2012, pp. 2899–2906.

Rodriguez, Olga, Michele L. Schaefer, Brock Wester, Yi-Chien Lee, Nathan Boggs, Howard A. Conner, Andrew C. Merkle, Stanley T. Fricke, Chris Albanese, and Vassilis E. Koliatsos, "Manganese-Enhanced Magnetic Resonance Imaging as a Diagnostic and Dispositional Tool After Mild-Moderate Blast Traumatic Brain Injury," *Journal of Neurotrauma*, Vol. 33, No. 7, April 2016, pp. 662–671.

Rubovitch, Vardit, Meital Ten-Bosch, Ofer Zohar, Catherine R. Harrison, Catherine Tempel-Brami, Elliot Stein, Barry J. Hoffer, Carey D. Balaban, Shaul Schreiber, Wen-Ta Chiu, and Chaim G. Pick, "A Mouse Model of Blast-Induced Mild Traumatic Brain Injury," *Experimental Neurology*, Vol. 232, No. 2, December 2011, pp. 280–289.

Ruff, Robert Louis, Ronald G. Riechers, 2nd, Xiao-Feng Wang, Traci Piero, and Suzanne Smith Ruff, "A Case-Control Study Examining Whether Neurological Deficits and PTSD in Combat Veterans Are Related to Episodes of Mild TBI," *BMJ Open*, Vol. 2, No. 2, 2012.

Ruff, Robert Louis, Suzanne Smith Ruff, and Xiao-Fend Wang, "Headaches Among Operation Iraqi Freedom/Operation Enduring Freedom Veterans with Mild Traumatic Brain Injury Associated with Exposures to Explosions," *Journal of Rehabilitation Research and Development*, Vol. 45, No. 7, 2008, pp. 941–952.

Säljö, Annette, Fredrik Arrhén, Hayde Bolouri, Maria Delen Solis Mayorga, and Anders C. Hamberger, "Neuropathology and Pressure in the Pig Brain Resulting from Low-Impulse Noise Exposure," *Journal of Neurotrauma*, Vol. 25, No. 12, December 2008, pp. 1397–1406.

Säljö, Annette, Hayde Bolouri, Maria Mayorga, Berndt Svensson, and Anders Hamberger, "Low-Level Blast Raises Intracranial Pressure and Impairs Cognitive Function in Rats: Prophylaxis with Processed Cereal Feed," *Journal of Neurotrauma*, Vol. 27, No. 2, February 2010, pp. 383–390.

Säljö, Annette, Maria Mayorga, Hayde Bolouri, Berndt Svensson, and Anders Hamberger, "Mechanisms and Pathophysiology of the Low-Level Blast Brain Injury in Animal Models," *NeuroImage*, Vol. 54, Supplement 1, January 2011, pp. S83–S88.

Saulle, Michael, and Brian D. Greenwald, "Chronic Traumatic Encephalopathy: A Review," *Rehabilitation Research and Practice*, Vol. 2012, 2012.

Saxena, Arpit, A. V. Ramesh, Poonam Raj Mehra, and D. K. Singh, "Short-Term Audiometric Profile in Army Recruits Following Rifle Firing: An Indian Perspective," *Indian Journal of Otology*, Vol. 22, No. 3, 2016, pp. 199–202.

Sayapathi, Balachandar S., Anselm Ting Su, and David Koh, "The Effectiveness of Applying Different Permissible Exposure Limits in Preserving the Hearing Threshold Level: A Systematic Review," *Journal of Occupational Health*, Vol. 56, No. 1, 2014, pp. 1–11.

Scheibel, Randall S., Mary R. Newsome, Maya Troyanskaya, Xiaodi D. Lin, Joel L. Steinberg, Majdi Radaideh, and Harvey S. Levin, "Altered Brain Activation in Military Personnel with One or More Traumatic Brain Injuries Following Blast," *Journal of the International Neuropsychological Society*, Vol. 18, No. 1, January 2012, pp. 89–100.

Scherer, Matthew R., and Michael C. Schubert, "Traumatic Brain Injury and Vestibular Pathology as a Comorbidity After Blast Exposure," *Physical Therapy*, Vol. 89, No. 9, September 2009, pp. 980–992.

Schindler, Abigail G., J. S. Meabon, Kathleen Farrell Pagulayan, R. Curtis Hendrickson, K. D. Meeker, M. Cline, G. Li, C. Sikkema, C. W. Wilkinson, and D. P. Perl, M. R. Rasking, E. R. Peskind, J. J. Clark, and D. G. Cook, "Blast–Related Disinhibition and Risk Seeking in Mice and Combat Veterans: Potential Role for Dysfunctional Phasic Dopamine Release," *Neurobiology of Disease*, Vol. 106, 2017, pp. 23–34.

Schulz, Theresa Y., "Troops Return with Alarming Rates of Hearing Loss," *Hearing Health*, Vol. 20, No. 3, 2004, pp. 18–21.

Schwab, Karen A., Gayle Baker, Brian J. Ivins, Melissa Sluss-Tiller, Warren E. Lux, and Deborah L. Warden, "The Brief Traumatic Brain Injury Screen (BTBIS): Investigating the Validity of a Self-Report Instrument for Detecting Traumatic Brain Injury (TBI) in Troops Returning from Deployment in Afghanistan and Iraq," *Neurology*, Vol. 66, No. 5, 2006, p. A235.

Shear, Deborah, *A Military-Relevant Model of Closed Concussive Head Injury: Longitudinal Studies Characterizing and Validating Single and Repetitive mTBI*, Tacoma, Wash.: Geneva Foundation, October 2014.

Sheffler, Julia L., Nicole C. Rushing, Ian H. Stanley, and Natalie J. Sachs-Ericsson, "The Long-Term Impact of Combat Exposure on Health, Interpersonal, and Economic Domains of Functioning," *Aging and Mental Health*, Vol. 20, No. 11, 2016, pp. 1202–1212.

Sherwood, Daniel, William E. Sponsel, Brian J. Lund, Walt Gray, Richard Watson, Sylvia L. Groth, Kimberly Thoe, Randolph D. Glickman, and Matthew A. Reilly, "Anatomical Manifestations of Primary Blast Ocular Trauma Observed in a Postmortem Porcine Model," *Investigative Ophthalmology and Visual Science*, Vol. 55, No. 2, February 2014, pp. 1124–1132.

Shinn-Cunningham, Barbara, Kenneth Grant, and Scott Bressler, *Diagnosing Contributions of Sensory and Cognitive Deficits to Hearing Dysfunction in Blast Exposed/TBI Service Members*, Boston, Mass.: Boston University, October 2016.

Shupak, Avi, Ilana Doweck, Dan Nachtigal, Orna Spitzer, and Carlos R. Gordon, "Vestibular and Audiometric Consequences of Blast Injury to the Ear," *Archives of Otolaryngology, Head and Neck Surgery*, Vol. 119, No. 12, December 1993, pp. 1362–1367.

Singh, Ajay K., Noah G. Ditkofsky, John D. York, Hani H. Abujudeh, Laura A. Avery, John F. Brunner, Aaron D. Sodickson, and Michael H. Lev, "Blast Injuries: From Improvised Explosive Device Blasts to the Boston Marathon Bombing," *Radiographics*, Vol. 36, No. 1, January–February 2016, pp. 295–307.

Song, Hailong, Jiankun Cui, Agnes Simonyi, Catherine E. Johnson, Graham K. Hubler, Ralph G. DePalma, and Zezong Gu, "Linking Blast Physics to Biological Outcomes in Mild Traumatic Brain Injury: Narrative Review and Preliminary Report of an Open-Field Blast Model," *Behavioural Brain Research*, Vol. 340, March 2016, pp. 147–158.

Stein, Murray B., and Thomas W. McAllister, "Exploring the Convergence of Posttraumatic Stress Disorder and Mild Traumatic Brain Injury," *American Journal of Psychiatry*, Vol. 166, No. 7, July 2009, pp. 768–776.

Stein, Thor D., Victor E. Alvarez, and Ann C. McKee, "Concussion in Chronic Traumatic Encephalopathy," *Current Pain Headache Reports*, Vol. 19, No. 10, October 2015.

Stephenson, Carol Merry, and Mark R. Stephenson, "Hearing Loss Prevention for Carpenters: Part 1—Using Health Communication and Health Promotion Models to Develop Training That Works," *Noise and Health*, Vol. 13, No. 51, 2011, pp. 113–121.

Stucken, Emily Z., and Robert S. Hong, "Noise-Induced Hearing Loss: An Occupational Medicine Perspective," *Current Opinion in Otolaryngology and Head and Neck Surgery*, Vol. 22, No. 5, October 2014, pp. 388–393.

Stuhmiller, James H., *A Health Hazard Assessment for Blast Overpressure Exposures*, San Diego, Calif.: Jaycor, 2003.

Svetlov, Stanislav, Ronald Hayes, and Olena Glushakova, *Molecular Signatures and Diagnostic Biomarkers of Cumulative Blast-Graded Mild TBI*, Alachua, Fla.: Banyan Biomarkers, December 2014.

Swe1234, "A Swedish Infantryman after a Long Day of Firing the Carl-Gustaf," Reddit post, April 29, 2015. As of July 18, 2019:
https://redd.it/34bx03

Tate, Charmaine M., Kevin K. W. Wang, Stephanie Eonta, Yang Zhang, Walter Carr, Frank C. Tortella, Ronald L. Hayes, and Gary H. Kamimori, "Serum Brain Biomarker Level, Neurocognitive Performance, and Self-Reported Symptom Changes in Soldiers Repeatedly Exposed to Low-Level Blast: A Breacher Pilot Study," *Journal of Neurotrauma*, Vol. 30, No. 19, October 2013, pp. 1620–1630.

Teland, Jan Arild, *Review of Blast Injury Prediction Models*, Kjeller, Norway: Norwegian Defence Research Establishment, March 14, 2012.

Tepe, Victoria, Christopher Smalt, Jeremy Nelson, Thomas Quatieri, and Kenneth Pitts, "Hidden Hearing Injury: The Emerging Science and Military Relevance of Cochlear Synaptopathy," *Military Medicine*, Vol. 182, No. 9–10, September 2017, pp. e1785–e1795.

Terrio, Heidi P., Lonnie A. Nelson, Lisa M. Betthauser, Jeri E. Harwood, and Lisa A. Brenner, "Postdeployment Traumatic Brain Injury Screening Questions: Sensitivity, Specificity, and Predictive Values in Returning Soldiers," *Rehabilitation Psychology*, Vol. 56, No. 1, 2011, pp. 26–31.

Thach, Allen B., Anthony J. Johnson, Robert B. Carroll, Ava Huchun, Darryl J. Ainbinder, Richard D. Stutzman, Sean M. Blaydon, Sheri L. DeMartelaere, Thomas H. Mader, Clifton S. Slade, Roger K. George, John P. Ritchey, Scott D. Barnes, and Lilia A. Fannin, "Severe Eye Injuries in the War in Iraq, 2003–2005," *Ophthalmology*, Vol. 115, No. 2, February 2008, pp. 377–382.

Thal, Serge C., and Winfried Neuhaus, "The Blood–Brain Barrier as a Target in Traumatic Brain Injury Treatment," *Archives of Medical Research*, Vol. 45, No. 8, November 2014, pp. 698–710.

Thiel, Kenneth J., Michael N. Dretsch, and William A. Ahroon, *The Effects of Low-Level Repetitive Blasts on Neuropsychological Functioning*, Fort Rucker, Ala.: U.S. Army Aeromedical Research Laboratory, December 2015. As of July 18, 2019:
http://www.dtic.mil/dtic/tr/fulltext/u2/1005243.pdf

Toyinbo, Peter A., Rodney D. Vanderploeg, Heather G. Belanger, Andrea M. Spehar, William A. Lapcevic, and Steven G. Scott, "A Systems Science Approach to Understanding Polytrauma and Blast-Related Injury: Bayesian Network Model of Data From a Survey of the Florida National Guard," *American Journal of Epidemiology*, Vol. 185, No. 2, January 2017, pp. 135–146.

Traumatic Brain Injury Task Force, *Report to the Surgeon General*, U.S. Army Medical Department, U.S. Army, May 15, 2007. As of July 18, 2019:
https://armymedicine.health.mil/reports

Tschiffely, Anna E., Stephen Thomas Ahlers, and Jacob N. Norris, "Examining the Relationship Between Blast-Induced Mild Traumatic Brain Injury and Posttraumatic Stress-Related Traits," *Journal of Neuroscience Research*, Vol. 93, No. 12, December 2015, pp. 1769–1777.

Tweedie, David, Lital Rachmany, Vardit Rubovitch, Yongqing Zhang, Kevin G. Becker, Evelyn Perez, Barry J. Hoffer, Chaim G. Pick, and Nigel H. Greig, "Changes in Mouse Cognition and Hippocampal Gene Expression Observed in a Mild Physical-and Blast-Traumatic Brain Injury," *Neurobiology of Disease*, Vol. 54, February 2013, pp. 1–11.

U.S. Government Accountability Office, *VA Health Care: Mild Traumatic Brain Injury Screening and Evaluation Implementation for OEF-OIF Veterans, but Challenges Remain*, Washington, D.C., GAO-08-276, February 2008.

U.S. Government Accountabilty Office, *Army Training: Efforts to Adjust Training Requirements Should Consider the Use of Virtual Training Devices*, report to congressional committees, Washington, D.C., GAO-16-636, 2016.

Valiyaveettil, Manojkumar, Yonas A. Alamneh, Stacy-Ann M. Miller, Rasha Hammamieh, Ying Wang, P. Bharath Arun, Yanling Wei, Samuel A. Oguntayo, and Madhusoodana P. Nambiar, "Preliminary Studies on Differential Expression of Auditory Functional Genes in the Brain After Repeated Blast Exposures," *Journal of Rehabilitation Research and Development*, Vol. 49, No. 7, 2012, pp. 1153–1162.

Valiyaveettil, Manojkumar, Yonas Alamneh, Ying Wang, Peethambaran Arun, Samuel Oguntayo, Yanling Wei, Joseph B. Long, and Madhusoodana P. Nambiar, "Contribution of Systemic Factors in the Pathophysiology of Repeated Blast-Induced Neurotrauma," *Neuroscience Letters*, Vol. 539, February 28, 2013.

Van Dyke, Sarah A., Bradley N. Axelrod, and Christian Schutte, "Test-Retest Reliability of the Traumatic Brain Injury Screening Instrument," *Military Medicine*, Vol. 175, No. 12, December 2010, pp. 947–949.

Vanderploeg, Rodney D., Heather G. Belanger, Ronnie D. Horner, Andrea M. Spehar, Gail Powell-Cope, Stephen L. Luther, and Steven G. Scott, "Health Outcomes Associated with Military Deployment: Mild Traumatic Brain Injury, Blast, Trauma, and Combat Associations in the Florida National Guard," *Archives of Physical Medicine and Rehabilitation*, Vol. 93, No. 11, November 2012, pp. 1887–1895.

Vargas, Bert B., and David W. Dodick, "Posttraumatic Headache," *Current Opinion in Neurology*, Vol. 25, No. 3, June 2012, pp. 284–289.

Valiyaveetil, Manojkumar, Yonas A. Alamneh, Stacy-Ann M. Miller, Ying Wang, P. Bharath Arun, Yanling Wei, Samuel A. Oguntayo, and Madhusoodana P. Nambiar, "Preliminary Studies on Differential Expression of Auditory Functional Genes in the Brain After Repeated Blast Exposures," *Journal of Rehabilitation Research and Development*, Vol. 49, No. 7, 2012, p. 1153–1162.

Walker, William C., Laura M. Franke, Adam P. Sima, and David X. Cifu, "Symptom Trajectories After Military Blast Exposure and the Influence of Mild Traumatic Brain Injury," *Journal of Head Trauma Rehabilitation*, Vol. 32, No. 3, May/June 2017, pp. e16–e26.

Walker, William C., *Epidemiological Study of Mild Traumatic Brain Injury Sequelae Caused by Blast Exposure During Operations Iraqi Freedom and Enduring Freedom*, Richmond, Va.: McGuire Research Institute, November 2014.

Wang, Ernest W., and Jason H. Huang, "Understanding and Treating Blast Traumatic Brain Injury in the Combat Theater," *Neurological Research*, Vol. 35, No. 3, April 2013, pp. 285–289.

Wares, Joanna R., Kathy W. Hoke, William Walker, Laura M. Franke, David X. Cifu, William Carne, and Cheryl Ford-Smith, "Characterizing Effects of Mild Traumatic Brain Injury and Posttraumatic Stress Disorder on Balance Impairments in Blast-Exposed Servicemembers and Veterans Using Computerized Posturography," *Journal of Rehabilitation Research and Development*, Vol. 52, No. 5, 2015, pp. 591–604.

Watson, Richard, Walt Gray, William E. Sponsel, Brian J. Lund, Randolph D. Glickman, Sylvia L. Groth, and Matthew A. Reilly, "Simulations of Porcine Eye Exposure to Primary Blast Insult," *Translational Vision Science and Technology*, Vol. 4, No. 4, August 2015.

Weichel, Eric D., Marcus H. Colyer, Spencer E. Ludlow, Kraig S. Bower, and Andrew S. Eiseman, "Combat Ocular Trauma Visual Outcomes During Operations Iraqi and Enduring Freedom," *Ophthalmology*, Vol. 115, No. 12, December 2008, pp. 2235–2245.

Wester, Brock, Michele Schaefer, Nathan Boggs, and Chris Bradburne, *The Importance of Neurogenic Inflammation in Blast-Induced Neurotrauma*, Laurel, Md.: Johns Hopkins University Applied Physics, 2014.

Wilk, Joshua E., Richard K. Herrell, Gary H. Wynn, Lyndon A. Riviere, and Charles W. Hoge, "Mild Traumatic Brain Injury (Concussion), Posttraumatic Stress Disorder, and Depression in U.S. Soldiers Involved in Combat Deployments: Association with Postdeployment Symptoms," *Psychosomatic Medicine*, Vol. 74, No. 3, April 2012, pp. 249–257.

Wilkinson, Charles W., *Blast Concussion mTBI, Hypopituitarism, and Psychological Health in OIF/OEF Veterans*, Seattle, Wash.: VA Puget Sound Health Care System, 2014.

Wilmington, Debra J., M. Samantha Lewis, Paula J. Myers, Frederick J. Gallun, and Stephen A. Fausti, "Hearing Impairment Among Soldiers: Special Considerations for Amputees," in Paul F. Pasquina and Rory A. Cooper, eds., *Care of the Combat Amputee*, Falls Church, Va., and Washington, D.C.: Office of the Surgeon General, Department of the Army, and Borden Institute, Walter Reed Army Medical Center, 2009.

Wiri, Suthee, Andrej Ritter, Jason M. Bailie, Cindy Needham, and Joshua L. Duckworth, "Computational Modeling of Blast Exposure Associated with Recoilless Weapons Combat Training," *Shock Waves*, Vol. 27, No. 6, November 2017, pp. 849–862.

Wong, Jessica M., Adam L. Halberstadt, Humberto A. Sainz, Kiran S. Mathews, Brian W. Chu, Laurel J. Ng, and Philemon C. Chan, "Mild Traumatic Brain Injury From Repeated Low-Level Blast Exposures," *ASME 2015 International Mechanical Engineering Congress and Exposition*, Vol. 3, 2015.

Woods, Amina S., Benoit Colsch, Shelley N. Jackson, Jeremy Post, Kathrine Baldwin, Aurelie Roux, Barry Hoffer, Brian M. Cox, Michael Hoffer, Vardit Rubovitch, Chaim G. Pick, J. Albert Schultz, and Carey Balaban, "Gangliosides and Ceramides Change in a Mouse Model of Blast Induced Traumatic Brain Injury," *ACS Chemical Neuroscience*, Vol. 4, No. 4, January 2013, pp. 594–600.

Xie, Kun, Hui Kuang, and Joe Z. Tsien, "Mild Blast Events Alter Anxiety, Memory, and Neural Activity Patterns in the Anterior Cingulate Cortex," *PLoS One*, Vol. 8, No. 5, May 2013, e64907.

Yaffe, Kristine, *Endophenotypes of Dementia Associated with Traumatic Brain Injury in Retired Military Personnel*, San Francisco, Calif. and Bethesda, Md.: Northern California Institute for Research and Education and Henry Jackson Foundation for the Advancement of Military Medicine, June 2015.

Yeh, Ping-Hong, Cheng Guan Koay, Binquan Wang, John Morissette, Elyssa Sham, Justin Senseney, David Joy, Alex Kubli, Chen-Haur Yeh, Victoria Eskay, Wei Liu, Louis M. French, Terrence R. Oakes, Gerard Riedy, and John Ollinger, "Compromised Neurocircuitry in Chronic Blast-Related Mild Traumatic Brain Injury," *Human Brain Mapping*, Vol. 38, No. 1, January 2017, pp. 352–369.

Zander, Nicole E., Thuvan Piehler, Mary E. Boggs, Rohan Banton, and Richard Benjamin, "In Vitro Studies of Primary Explosive Blast Loading on Neurons," *Journal of Neuroscience Research*, Vol. 93, No. 9, September 2015, pp. 1353–1363.

Zhang, Jinsheng, Anthony Cacace, E. Mark Haacke, Bruce Berkowitz, Jiani Hu, Randall Benson, and John Woodard, *Parallel Human and Animal Models of Blast- and Concussion-Induced Tinnitus and Related Traumatic Brain Injury (TBI)*, Detroit, Mich.: Wayne State University, 2014.

Zhu, Feng, Clifford C. Chou, King H. Yang, and Albert I. King, "A Theoretical Analysis of Stress Wave Propagation in the Head Under Primary Blast Loading," *Proceedings of the Institution of Mechanical Engineers, Part H: Journal of Engineering in Medicine*, Vol. 228, No. 5, 2014, pp. 439–445.

Zollman, Felise S., Christine Starr, Bethany Kondiles, Cherina Cyborski, and Eric B. Larson, "The Rehabilitation Institute of Chicago Military Traumatic Brain Injury Screening Instrument: Determination of Sensitivity, Specificity, and Predictive Value," *Journal of Head Trauma Rehabilitation*, Vol. 29, No. 1, January–February 2014, pp. 99–107.

# References

Ahlers, Stephen Thomas, Elaina Vasserman-Stokes, Michael Christopher Shaughness, Aaron Andrew Hall, Debra Ann Shear, Mikulas Chavko, Richard Michael McCarron, and James Radford Stone, "Assessment of the Effects of Acute and Repeated Exposure to Blast Overpressure in Rodents: Toward a Greater Understanding of Blast and the Potential Ramifications for Injury in Humans Exposed to Blast," *Frontiers in Neurology*, Vol. 3, Article 32, March 2012.

Baalman, Kelli L., R. James Cotton, S. Neil Rasband, and Matthew N. Rasband, "Blast Wave Exposure Impairs Memory and Decreases Axon Initial Segment Length," *Journal of Neurotrauma*, Vol. 30, No. 9, May 2013, pp. 741–751.

"Blast Overpressure Is Generated from the Firing of Weapons, and May Cause Brain Injury," *ScienceDaily*, January 20, 2009. As of July 18, 2019:
https://www.sciencedaily.com/releases/2009/01/090119091112.htm

Blennow, Kaj, Michael Jonsson, Niels Peter Andreasen, Lars Erik Rosengren, Anders Wallin, Pekka A. Hellström, and Henrik Zetterber, "No Neurochemical Evidence of Brain Injury After Blast Overpressure by Repeated Explosions or Firing Heavy Weapon," *Acta Neurologica Scandinavica*, Vol. 123, No. 4, April 2011, pp. 245–251.

Budde, Matthew D., Alok Shah, Michael McCrea, William E. Cullinan, Frank A. Pintar, and Brian D. Stemper, "Primary Blast Traumatic Brain Injury in the Rat: Relating Diffusion Tensor Imaging and Behavior," *Frontiers in Neurology*, Vol. 4, Article 154, October 2013.

Carr, Walter, Elena Polejaeva, Anna Grome, Beth Crandall, Christina LaValle, Stephanie Eonta, and Lee Young, "Relation of Repeated Low-Level Blast Exposure with Symptomology Similar to Concussion," *Journal of Head Trauma Rehabilitation*, Vol. 30, No. 1, January–February 2015, pp. 47–55.

Carr, Walter, James R. Stone, Tim Walilko, Lee Ann Young, Tianlu Li Snook, Michelle E. Paggi, Jack W. Tsao, Christopher J. Jankosky, Robert V. Parish, and Stephen T. Ahlers, "Repeated Low-Level Blast Exposure: A Descriptive Human Subjects Study," *Military Medicine*, Vol. 181, Supplement 5, May 2016, pp. 28–39.

Chavko, Mikulas, Tomas Watanabe, Saleena Adeeb, Jason Lankasky, Stephen T. Ahlers, and Richard M. McCarron, "Relationship Between Orientation to a Blast and Pressure Wave Propagation Inside the Rat Brain," *Journal of Neuroscience Methods*, Vol. 195, No. 1, January 2011, pp. 61–66.

Coats, Brittany, and Daniel Shedd, *Temporal Progression of Visual Injury from Blast Exposure*, Salt Lake City, Utah: University of Utah, 2016. As of July 18, 2019:
http://www.dtic.mil/dtic/tr/fulltext/u2/1020693.pdf

Committee on Gulf War and Health, *Gulf War and Health*, Vol. 7: *Long-Term Consequences of Traumatic Brain Injury*, Washington, D.C.: The National Academies Press, 2008.

Courtney, Michael W., and Amy C. Courtney, "Working Toward Exposure Thresholds for Blast-Induced Traumatic Brain Injury: Thoracic and Acceleration Mechanisms," *NeuroImage*, Vol. 54, Supplement 1, January 2011, pp. S55–S61.

Department of Defense Blast Injury Research Program Coordinating Office, *Biomedical Basis for Mild Traumatic Brain Injury (mTBI) Environmental Sensor Threshold Values*, McLean, Va., 2014. As of July 18, 2019:
https://blastinjuryresearch.amedd.army.mil/assets/docs/sos/meeting_proceedings/2014_SoS_Meeting_Proceedings.pdf

Dougherty, Amber L., Andrew J. MacGregor, Peggy P. Han, Erik S. Viirre, Kevin J. Heltemes, and Michael R. Galarneau, "Blast-Related Ear Injuries Among U.S. Military Personnel," *Journal of Rehabilitation Research and Development*, Vol. 50, No. 6, November 2013, pp. 893–904.

Elder, Gregory A., Nathan P. Dorr, Rita De Gasperi, Miguel A. Gama Sosa, Michael C. Shaughness, Eric Maudlin-Jeronimo, Aaron A. Hall, Richard M. McCarron, and Stephen T. Ahlers, "Blast Exposure Induces Post-Traumatic Stress Disorder-Related Traits in a Rat Model of Mild Traumatic Brain Injury," *Journal of Neurotrauma*, Vol. 29, No. 16, 2012, pp. 2564–2575.

Elder, Gregory A., James R. Stone, and Stephen T. Ahlers, "Effects of Low-Level Blast Exposure on the Nervous System: Is There Really a Controversy?" *Frontiers in Neurology*, Vol. 5, Article 269, December 2014.

Ewert, Donald L., Jianzhong Lu, Wei Li, Xiaoping Du, Robert Floyd, and Richard Kopke, "Antioxidant Treatment Reduces Blast-Induced Cochlear Damage and Hearing Loss," *Hearing Research*, Vol. 285, No. 1–2, March 2012, pp. 29–39.

Fausti, Stephen A., Debra J. Wilmington, Frederick J. Gallun, Paula J. Myers, and James A. Henry, "Auditory and Vestibular Dysfunction Associated with Blast-Related Traumatic Brain Injury," *Journal of Rehabilitation Research and Development*, Vol. 46, No. 6, 2009, pp. 797–809.

Fischer, Barbara L., Michael Parsons, Sally Durgerian, Christine Reece, Lyla Mourany, Mark J. Lowe, Erik B. Beall, Katherine A. Koenig, Stephen E. Jones, Mary R. Newsome, Randall S. Scheibel, Elisabeth A. Wilde, Maya Troyanskaya, Tricia L. Merkley, Mark Walker, Harvey S. Levin, and Stephen M. Rao, "Neural Activation During Response Inhibition Differentiates Blast from Mechanical Causes of Mild to Moderate Traumatic Brain Injury," *Journal of Neurotrauma*, Vol. 31, No. 2, 2014, pp. 169–179.

Fish, Lauren, and Paul Scharre, *Protecting Warfighters from Blast Injury*, Washington, D.C.: Center for a New American Security, 2018.

Gama Sosa, Miguel A., Rita De Gasperi, Pierce L. Janssen, Frank J. Yuk, Pameka C. Anazodo, Paul E. Pricop, Alejandro J. Paulino, Bridget A. Wicinski, Michael Christopher Shaughness, Eric Maudlin-Jeronimo, Aaron A. Hall, Dara L. Dickstein, Richard M. McCarron, Mikulas Chavko, Patrick R. Hof, Stephen T. Ahlers, and Gregory A. Elder, "Selective Vulnerability of the Cerebral Vasculature to Blast Injury in a Rat Model of Mild Traumatic Brain Injury," *Acta Neuropathologica Communications*, Vol. 2, No. 67, 2014.

Gama Sosa, Miguel A., Rita De Gasperi, Alejandro J. Paulino, Paul E. Pricop, Michael C. Shaughness, Eric Maudlin-Jeronimo, Aaron A. Hall, William G. M. Janssen, Frank J. Yuk, Nathan P. Dorr, Dara L. Dickstein, Richard M. McCarron, Mikulas Chavko, Patrick R. Hof, Stephen T. Ahlers, and Gregory A. Elder, "Blast Overpressure Induces Shear-Related Injuries in the Brain of Rats Exposed to a Mild Traumatic Brain Injury," *Acta Neuropathologica Communications*, Vol. 1, No. 51, 2013.

Gan, Rong, *Biomechanical Modeling and Measurement of Blast Injury and Hearing Protection Mechanisms*, Norman, Okla.: University of Oklahoma, Norman, 2015.

Ganpule, Shailesh G., Linxia Gu, Aaron Alai, and Namas Chandra, "Role of Helmet in the Mechanics of Shock Wave Propagation Under Blast Loading Conditions," *Computer Methods in Biomechanics and Biomedical Engineering*, Vol. 15, No. 11, 2012, pp. 1233–1244.

Goverover, Yael, and Nancy Chiaravalloti, "The Impact of Self-Awareness and Depression on Subjective Reports of Memory, Quality-of-Life and Satisfaction with Life Following TBI," *Brain Injury*, Vol. 28, No. 2, 2014, pp. 174–180.

Grujicic, M. Danica, W. C. Bell, B. Pandurangan, and T. He, "Blast-Wave Impact-Mitigation Capability of Polyurea When Used as Helmet Suspension-Pad Material," *Materials and Design*, Vol. 31, No. 9, 2010, pp. 4050–4065.

Grujicic, M. Danica, William Cameron Bell, Bhaskar Pandurangan, and Patrick Glomski, "Fluid/Structure Interaction Computational Investigation of Blast-Wave Mitigation Efficacy of the Advanced Combat Helmet," *Journal of Materials Engineering and Performance*, Vol. 20, No. 6, August 2011, pp. 877–893.

Hamilton, Jon, "Do U.S. Troops Risk Brain Injury When They Fire Heavy Weapons?" *National Public Radio*, April 5, 2017. As of October 1, 2017:
https://www.npr.org/sections/health-shots/2017/04/05/522613294/
do-u-s-troops-risk-brain-injury-when-they-fire-heavy-weapons

Haran, F. J., Aimee L. Alphonso, Alia Creason, Justin S. Campbell, Dagny Johnson, Emily Young, and Jack W. Tsao, "Analysis of Post-Deployment Cognitive Performance and Symptom Recovery in U.S. Marines," *PLoS One*, Vol. 8, No. 11, 2013, e79595.

Heldt, Scott A., Andrea J. Elberger, Yunping Deng, Natalie Hart Guley, Nobel A. Del Mar, Joshua T. Rogers, Gy Won Choi, Jessica Ferrell, Tonia S. Rex, Marcia G. Honig, and Anton Reiner, "A Novel Closed-Head Model of Mild Traumatic Brain Injury Caused by Primary Overpressure Blast to the Cranium Produces Sustained Emotional Deficits in Mice," *Frontiers in Neurology*, Vol. 5, Article 2, 2014.

Helling, Eric Robert, "Otologic Blast Injuries Due to the Kenya Embassy Bombing," *Military Medicine*, Vol. 169, No. 11, November 2004, pp. 872–876.

Hoge, Charles W., Dennis McGurk, Jeffrey L. Thomas, Anthony L. Cox, Charles C. Engel, and Carl A. Castro, "Mild Traumatic Brain Injury in U.S. Soldiers Returning from Iraq," *New England Journal of Medicine*, Vol. 358, No. 5, January 2008, pp. 453–463.

Huber, Bertrand R., James S. Meabon, Zachary S. Hoffer, Jing Zhang, Jake G. Hoekstra, Kathleen F. Pagulayan, Pamela J. McMillan, Cynthia L. Mayer, William A. Banks, Brian C. Kraemer, Murray A. Raskind, Dorian B. McGavern, Elaine R. Peskind, and David G. Cook, "Blast Exposure Causes Dynamic Microglial/Macrophage Responses and Microdomains of Brain Microvessel Dysfunction," *Neuroscience*, Vol. 319, 2016, pp. 206–220.

Job, Agnes, Pascal Hamery, Sebastien De Mezzo, J-C Fialaire, Andre A. le Roux, Micahel Untereiner, Federica Cardinale, Hugues Michel, Celine Klein, and Bill Belcourt, "Rifle Impulse Noise Affects Middle-Ear Compliance in Soldiers Wearing Protective Earplugs," *International Journal of Audiology*, Vol. 55, No. 1, 2016, pp. 30–37.

Kamnaksh, Alaa, Matthew D. Budde, Erzsebet Kovesdi, Joseph B. Long, Joseph A. Frank, and Denes V. Agoston, "Diffusion Tensor Imaging Reveals Acute Subcortical Changes After Mild Blast-Induced Traumatic Brain Injury," *Scientific Reports*, Vol. 4, 2014.

Karr, Justin E., Corson N. Areshenkoff, and Mauricio A. Garcia-Barrera, "The Neuropsychological Outcomes of Concussion: A Systematic Review of Meta-Analyses on the Cognitive Sequelae of Mild Traumatic Brain Injury," *Neuropsychology*, Vol. 28, No. 3, May 2014, pp. 321–336.

Kulkarni, Shilpa G., Xiang Gao, Stephen V. Horner, Ji Qing Zheng, and N. V. David, "Ballistic Helmets—Their Design, Materials, and Performance Against Traumatic Brain Injury," *Composite Structures*, Vol. 101, July 2013, pp. 313–331.

Luo, Jian, Andy Nguyen, Saul Villeda, Hui Zhang, Zhaoqing Ding, Derek Lindsey, Gregor Bieri, Joseph M. Castellano, Gary S. Beaupre, and Tony Wyss-Coray, "Long-Term Cognitive Impairments and Pathological Alterations in a Mouse Model of Repetitive Mild Traumatic Brain Injury," *Frontiers in Neurology*, Vol. 5, Article 12, February 2014.

Mac Donald, Christine L., Ann M. Johnson, Dana Cooper, Elliot C. Nelson, Nicole J. Werner, Joshua S. Shimony, Abraham Z. Snyder, Marcus E. Raichle, John R. Witherow, Raymond Fang, Stephen F. Flaherty, and David L. Brody, "Detection of Blast-Related Traumatic Brain Injury in U.S. Military Personnel," *New England Journal of Medicine*, Vol. 364, No. 22, June 2011, pp. 2091–2100.

Macera, Caroline A., Hilary Jeanne Aralis, Andrew J. MacGregor, Mitchell J. Rauh, and Michael R. Galarneau, "Postdeployment Symptom Changes and Traumatic Brain Injury and/or Posttraumatic Stress Disorder in Men," *Journal of Rehabilitation Research and Development*, Vol. 49, No. 8, 2012, pp. 1197–1208.

Management of Concussion-mild Traumatic Brain Injury Working Group, *VA/DoD Clinical Practice Guideline for Management of Concussion-Mild Traumatic Brain Injury*, Washington, D.C.: U.S. Department of Veterans Affairs and U.S. Department of Defense, 2016. As of July 18, 2019: https://www.healthquality.va.gov/guidelines/Rehab/mtbi/

McEntire, B. Joseph, V. Carol Chancey, Timothy Walilko, Gregory T. Rule, Gregory Weiss, Cameron Bass, and Jay Shridharani, *Helmet Sensor-Transfer Function and Model Development*, San Antonio, Tex.: True Research Foundation, 2010.

Miller, Greg, "Blast Injuries Linked to Neurodegeneration in Veterans," *Science*, Vol. 336, No. 6083, May 2012, pp. 790–791.

Moochhala, Shabbir M., Shirhan Md, Jia Lu, Choo-Hua Teng, and Colin Greengrass, "Neuroprotective Role of Aminoguanidine in Behavioral Changes After Blast Injury," *Journal of Trauma and Acute Care Surgery*, Vol. 56, No. 2, March 2004, pp. 393–403.

Moss, William C., Michael J. King, and Eric G. Blackman, "Skull Flexure from Blast Waves: A Mechanism for Brain Injury with Implications for Helmet Design," *Physical Review Letters*, Vol. 103, No. 10, September 2009.

Nyein, Michelle K., Amanda M. Jason, Li Yu, Claudio M. Pita, John D. Joannopoulos, David F. Moore, and Raul A. Radovitzky, "In Silico Investigation of Intracranial Blast Mitigation with Relevance to Military Traumatic Brain Injury," *Proceedings of the National Academy of Sciences*, Vol. 107, No. 48, October 2010, pp. 20703–20708.

Oishi, Naoki, and Jochen Schacht, "Emerging Treatments for Noise-Induced Hearing Loss," *Expert Opinion on Emerging Drugs*, Vol. 16, No. 2, June 2011, pp. 235–245.

Okpala, Nnaemeka, "Knowledge and Attitude of Infantry Soldiers to Hearing Conservation," *Military Medicine*, Vol. 172, No. 5, May 2007, pp. 520–522.

Panzer, Matthew B., Cameron R. Dale Bass, Karin A. Rafaels, Jay Shridharani, and Bruce P. Capehart, "Primary Blast Survival and Injury Risk Assessment for Repeated Blast Exposures," *Journal of Trauma and Acute Care Surgery*, Vol. 72, No. 2, February 2012, pp. 454–466.

Parish, R., W. Carr, Michelle Paggi, V. Anderson-Barnes, and M. Kelly, "The Neurocognitive Effect of Exposure to Repeated Low-Level Blasts in a Military Sample," *Archives of Clinical Neuropsychology*, Vol. 24, No. 5, August 2009, pp. 503–504.

Park, Eugene, James J. Gottlieb, Bob Cheung, Pang Nin Shek, and Andrew James Baker, "A Model of Low-Level Primary Blast Brain Trauma Results in Cytoskeletal Proteolysis and Chronic Functional Impairment in the Absence of Lung Barotrauma," *Journal of Neurotrauma*, Vol. 28, No. 3, March 2011, pp. 343–357.

Perez-Garcia, Georgina, Rita De Gasperi, Miguel A. Gama Sosa, Gissel M. Perez, Alena Otero-Pagan, Anna Tschiffely, Richard M. McCarron, Stephen T. Ahlers, Gregory A. Elder, and Sam Gandy, "PTSD-Related Behavioral Traits in a Rat Model of Blast-Induced mTBI Are Reversed by the mGluR2/3 Receptor Antagonist BCI-838," *eNeuro*, Vol. 5, No. 1, January/February 2018.

Perez-Garcia, Georgina, Miguel A. Gama Sosa, Rita De Gasperi, Margaret Lashof-Sullivan, Eric Maudlin-Jeronimo, James R. Stone, Fatemeh Mohammadian Haghighi, Stephen Thomas Ahlers, and Gregory A. Elder, "Exposure to a Predator Scent Induces Chronic Behavioral Changes in Rats Previously Exposed to Low-Level Blast: Implications for the Relationship of Blast-Related TBI to PTSD," *Frontiers in Neurology*, Vol. 7, Article 176, 2016.

Por, Elaine D., Jae-Hyek Choi, and Brian J. Lund, "Low-Level Blast Exposure Increases Transient Receptor Potential Vanilloid 1 (TRPV1) Expression in the Rat Cornea," *Current Eye Research*, Vol. 41, No. 10, 2016, pp. 1294–1301.

Pun, Pamela Boon Li, Mary Kan, Agus Salim, Zhaohui Li, Kian Chye Ng, Shabbir M. Moochhala, Eng-Ang Ling, Mui Hong Tan, and Jia Lu, "Low Level Primary Blast Injury in Rodent Brain," *Frontiers in Neurology*, Vol. 2, Article 19, April 2011.

Reid, Matthew W., Kelly J. Miller, Rael T. Lange, Douglas B. Cooper, David F. Tate, Jason Bailie, Tracey A. Brickell, Louis M. French, Sarah Asmussen, and Jan E. Kennedy, "A Multisite Study of the Relationships Between Blast Exposures and Symptom Reporting in a Post-Deployment Active Duty Military Population with Mild Traumatic Brain Injury," *Journal of Neurotrauma*, Vol. 31, No. 23, December 2014, pp. 1899–1906.

Rodriguez, Olga, Michele L. Schaefer, Brock Wester, Yi-Chien Lee, Nathan Boggs, Howard A. Conner, Andrew C. Merkle, Stanley T. Fricke, Chris Albanese, and Vassilis E. Koliatsos, "Manganese-Enhanced Magnetic Resonance Imaging as a Diagnostic and Dispositional Tool After Mild-Moderate Blast Traumatic Brain Injury," *Journal of Neurotrauma*, Vol. 33, No. 7, April 2016, pp. 662–671.

Rubovitch, Vardit, Meital Ten-Bosch, Ofer Zohar, Catherine R. Harrison, Catherine Tempel-Brami, Elliot Stein, Barry J. Hoffer, Carey D. Balaban, Shaul Schreiber, Wen-Ta Chiu, and Chaim G. Pick, "A Mouse Model of Blast-Induced Mild Traumatic Brain Injury," *Experimental Neurology*, Vol. 232, No. 2, December 2011, pp. 280–289.

Säljö, Annette, Fredrik Arrhén, Hayde Bolouri, Maria Delen Solis Mayorga, and Anders C. Hamberger, "Neuropathology and Pressure in the Pig Brain Resulting from Low-Impulse Noise Exposure," *Journal of Neurotrauma*, Vol. 25, No. 12, December 2008, pp. 1397–1406.

Säljö, Annette, Hayde Bolouri, Maria Mayorga, Berndt Svensson, and Anders Hamberger, "Low-Level Blast Raises Intracranial Pressure and Impairs Cognitive Function in Rats: Prophylaxis with Processed Cereal Feed," *Journal of Neurotrauma*, Vol. 27, No. 2, February 2010, pp. 383–390.

Säljö, Annette, Maria Mayorga, Hayde Bolouri, Berndt Svensson, and Anders Hamberger, "Mechanisms and Pathophysiology of the Low-Level Blast Brain Injury in Animal Models," *NeuroImage*, Vol. 54, Supplement 1, January 2011, pp. S83–S88.

Schulz, Theresa Y., "Troops Return with Alarming Rates of Hearing Loss," *Hearing Health*, Vol. 20, No. 3, 2004, pp. 18–21.

Shupak, Avi, Ilana Doweck, Dan Nachtigal, Orna Spitzer, and Carlos R. Gordon, "Vestibular and Audiometric Consequences of Blast Injury to the Ear," *Archives of Otolaryngology, Head and Neck Surgery*, Vol. 119, No. 12, December 1993, pp. 1362–1367.

Stephenson, Carol Merry, and Mark R. Stephenson, "Hearing Loss Prevention for Carpenters: Part 1—Using Health Communication and Health Promotion Models to Develop Training That Works," *Noise and Health*, Vol. 13, No. 51, 2011, pp. 113–121.

Tate, Charmaine M., Kevin K. W. Wang, Stephanie Eonta, Yang Zhang, Walter Carr, Frank C. Tortella, Ronald L. Hayes, and Gary H. Kamimori, "Serum Brain Biomarker Level, Neurocognitive Performance, and Self-Reported Symptom Changes in Soldiers Repeatedly Exposed to Low-Level Blast: A Breacher Pilot Study," *Journal of Neurotrauma*, Vol. 30, No. 19, October 2013, pp. 1620–1630.

Teland, Jan Arild, *Review of Blast Injury Prediction Models*, Kjeller, Norway: Norwegian Defence Research Establishment, March 14, 2012.

Thiel, Kenneth J., Michael N. Dretsch, and William A. Ahroon, *The Effects of Low-Level Repetitive Blasts on Neuropsychological Functioning*, Fort Rucker, Ala.: U.S. Army Aeromedical Research Laboratory, December 2015. As of July 18, 2019:
https://pdfs.semanticscholar.org/3adf/2ed20a438d600950332514cf8aa6c5dc5f09.pdf

Toyinbo, Peter A., Rodney D. Vanderploeg, Heather G. Belanger, Andrea M. Spehar, William A. Lapcevic, and Steven G. Scott, "A Systems Science Approach to Understanding Polytrauma and Blast-Related Injury: Bayesian Network Model of Data From a Survey of the Florida National Guard," *American Journal of Epidemiology*, Vol. 185, No. 2, January 2017, pp. 135–146.

Tweedie, David, Lital Rachmany, Vardit Rubovitch, Yongqing Zhang, Kevin G. Becker, Evelyn Perez, Barry J. Hoffer, Chaim G. Pick, and Nigel H. Greig, "Changes in Mouse Cognition and Hippocampal Gene Expression Observed in a Mild Physical-and Blast-Traumatic Brain Injury," *Neurobiology of Disease*, Vol. 54, February 2013, pp. 1–11.

Walker, William C., Laura M. Franke, Adam P. Sima, and David X. Cifu, "Symptom Trajectories After Military Blast Exposure and the Influence of Mild Traumatic Brain Injury," *Journal of Head Trauma Rehabilitation*, Vol. 32, No. 3, May/June 2017, pp. e16–e26.

Wilk, Joshua E., Richard K. Herrell, Gary H. Wynn, Lyndon A. Riviere, and Charles W. Hoge, "Mild Traumatic Brain Injury (Concussion), Posttraumatic Stress Disorder, and Depression in U.S. Soldiers Involved in Combat Deployments: Association with Postdeployment Symptoms," *Psychosomatic Medicine*, Vol. 74, No. 3, April 2012, pp. 249–257.

Wilmington, Debra J., M. Samantha Lewis, Paula J. Myers, Frederick J. Gallun, and Stephen A. Fausti, "Hearing Impairment Among Soldiers: Special Considerations for Amputees," in Paul F. Pasquina and Rory A. Cooper, eds., *Care of the Combat Amputee*, Falls Church, Va., and Washington, D.C.: Office of the Surgeon General, Department of the Army, and Borden Institute, Walter Reed Army Medical Center, 2009.

Wiri, Suthee, Andrej Ritter, Jason M. Bailie, Cindy Needham, and Joshua L. Duckworth, "Computational Modeling of Blast Exposure Associated with Recoilless Weapons Combat Training," *Shock Waves*, Vol. 27, No. 6, November 2017, pp. 849–862.

Wojcik, Barbara E., C. R. Stein, Karen A. Bagg, Rebecca J. Humphrey, and Jason Orosco, "Traumatic Brain Injury Hospitalizations of U.S. Army Soldiers Deployed to Afghanistan and Iraq," *American Journal of Preventive Medicine*, Vol. 38, No. 1, January 2010, pp. S108–S116.

Woods, Amina S., Benoit Colsch, Shelley N. Jackson, Jeremy Post, Kathrine Baldwin, Aurelie Roux, Barry Hoffer, Brian M. Cox, Michael Hoffer, Vardit Rubovitch, Chaim G. Pick, J. Albert Schultz, and Carey Balaban, "Gangliosides and Ceramides Change in a Mouse Model of Blast Induced Traumatic Brain Injury," *ACS Chemical Neuroscience*, Vol. 4, No. 4, January 2013, pp. 594–600.

Xie, Kun, Hui Kuang, and Joe Z. Tsien, "Mild Blast Events Alter Anxiety, Memory, and Neural Activity Patterns in the Anterior Cingulate Cortex," *PLoS One*, Vol. 8, No. 5, May 2013, e64907.